GOODWILL'S

AWESOME EXPE

CW01455541

FORCE
&
MOTION

Michael Dispezio

Illustrations by Catherine Leary

GOODWILL PUBLISHING HOUSE®
B-3 RATTAN JYOTI, 18 RAJENDRA PLACE
NEW DELHI-110008 (INDIA)

© Michael Dispezio

All Rights Reserved

This special low priced Indian reprint is published by arrangement with Sterling Publishing Company, Inc. New York, U.S.A.

No part of this publication may be reproduced, stored in a retrieval system, or transmitted in any form or by any means, electronic, mechanical, photocopying, recording or otherwise, without the prior written permission of the publisher.

Published in India by:
GOODWILL PUBLISHING HOUSE®
B-3 Rattan Jyoti, 18 Rajendra Place
New Delhi–110 008 (INDIA)
Tel.: 25820556, 25750801, 25755559
Fax: 91-11-25764396
E-mail: goodwillpub@vsnl.net
           ylp@bol.net.in
Website: www.goodwillpublishinghouse.com

Printed at :-
B.B. Press
Delhi

CONTENTS

PART THREE
WHAT'S THE ATTRACTION?

PART FOUR
THE PUSHES AND PULLS OF AIR

PART FIVE
MOTION MADNESS

SAFETY FIRST

Follow all instructions, cautions, and safety notes. To protect your eyes, wear safety goggles when performing all of the experiments. Conduct every experiment with proper supervision. Have an adult perform all steps that use a flame, wall outlet, sharp point, cutting edge, or any other potentially dangerous tool. Neither the author nor publisher shall be liable for injuries that may be caused by not following the experiment steps or adhering to the safety notes.

INTRODUCTION

This book is a guide. Its primary purpose is to accompany you through more than seventy adventures in learning. As you perform these experiments, you'll celebrate the magic of science. You'll also observe how science isn't a distant notion limited to classrooms, laboratories, books, and PBS specials. Science is all around you!

Unlike many other subjects, science is constructed from inquiry. This philosophy of exploration is a cornerstone of the National Science Education Standards. It is also the premise upon which the *Awesome Experiments in Science* series has been created. With an increased focus on understanding (and NOT memorizing facts), these books offer kid-friendly experiments that will engage, harness, and nurture your thinking skills.

PART ONE

STAYING PUT

1.1 LOOP LAUNCH

Do you like playing science tricks on people? If so, this first adventure may be the perfect challenge to present to your friends. There's only one problem. You'll need to figure out how to solve it!

All you have to do is knock out the paper loop so that the coin falls directly into the cup below.

Materials
* 1 inch × 11 inch paper strip
* tape
* coin
* cup

To Do
Tape the ends of the paper strip together in order to form a large loop. Position it on the mouth of a cup. Carefully balance a coin on the uppermost part of the loop.

Now that you've assembled the setup, you need to knock out the loop with a single swiping motion so that the coin drops into the cup. It may sound easy, but just give it a try!

The Science of Failure
If you strike the outside of the paper loop, you compress the flexible shape. The top of the loop shoots upwards. Anything positioned on the top (such as your coin) is launched into the air and away from the target landing zone.

The Science of Success
To make this work, your strike must be directed on the inner side of the loop. As your hand moves, it misses the outer loop but "grabs" the inner surface of the opposite side. As your strike contacts the loop, it flattens the shape and drags it off the cup. Since the coin is not launched into the air, it drops straight down into the cup below.

Presentation Hint

When you perform this trick in front of others, don't tell them your secret. With enough practice, your hand will move so quickly that no one will see you strike the inner surface of the opposite side of the loop. They will think you're striking the outside of the loop.

✓

1.2 UNSEEN GLUE

Have you ever watched a magician yank away a tablecloth to leave the setting of glasses, plates, and silverware untouched? Perhaps you wondered if some unseen magnet or sticky surface helped keep the objects stationary. Probably not. For most performers, this trick depends only upon a basic knowledge of forces and motion.

Materials

* *empty plastic pop container (with cap)*
* *strip of cloth*

To Do

Fill the pop container with water and secure its cap. Dry off the container and make sure that there are no leaks. Place a strip of cloth on top of a flat surface, such as a desk or a table. Position the water-filled container on the cloth.

Grab hold of the edge of the cloth. With a firm "snap," pull out the cloth. As you pull the material, keep you arm at the same level as the table top.

Once you've learned this technique, remove the water from the container and try it again. Is it easier or harder to perform this trick when the container is empty? Can you explain why?

The Science

This trick is based upon something that scientists call *inertia* (IN-ur-sha). Inertia is the resistance to change. All objects have inertia. Things at rest tend to stay at rest. Things in motion tend to stay in motion. The inertia of the water-filled container kept it in the same spot as the cloth was pulled out from below it.

The more massive an object is, the more inertia it has. The container filled with water had more mass; therefore, it had more inertia. Since it was more resistant to changing, this water-filled container was easier to keep in one place than the empty container.

1.3 CARD SHOT

Although the tablecloth trick is a great show stopper, it's difficult to carry with you. This trick, however, uses pocket-sized props to show off the same principle.

Materials
* coin
* plastic playing card or an index card

To Do
Turn up the palm of one of your hands. Make a fist with that hand, but extend your index finger. Balance a playing card on the tip of the extended finger. Place a coin in the center of the card so that it too balances. Using a finger from the other hand, flick out the card. If your flick remains level, only the card will sail away. The coin will remain balanced on your fingertip.

The Science
All things, including coins, have inertia. When you flicked your finger, the forward motion was transferred to the card—not the coin. The card sailed off. Since the coin wasn't struck, its inertia helped keep it in place.

CHECK IT OUT! Suppose the card's surface wasn't slick. Suppose you glued sand grains or taped sandpaper to its surface. How might this affect the movement of the coin? Make a guess and then test your prediction.

1.4 STATIONARY SPIN

So far, you've seen that solid objects, like coins and pop bottles, resist changes in their movement. But did you know that liquids and other fluids also have inertia? Here's a setup that shows how a liquid tends to stay put.

Materials
* *mug*
* *water*

* *vegetable oil*
* *food coloring*

To Do
Fill a mug halfway with water. Carefully pour oil into the cup so that the surface of the water becomes covered by a thin layer of oil. Place four drops of food coloring onto the oil. The drops should be positioned at the corners of an imaginary square as shown in the illustration.

While the mug remains on the table, grasp it from above. Quickly turn your wrist so that the mug spins about a quarter turn. What happens to the droplets? Can you explain your observations?

The Science
Oil and water don't mix. When the oil was added to the mug, it floated upon the water's surface. The drops of food coloring entered the oil surroundings but did not flow into the water layer below. Since food coloring does not mix with oil, the dye remained as round droplets.

Although the mug spun through one quarter turn, the drops (and the mug's other liquids) remained mostly stationary. It was the inertia of these substances that kept them from moving.

1.5 EGG SPIN

"s it fresh or hard-boiled?"
"How am I supposed to know? Do I look like a psychic or a cook?"
"What are you, then?"
"I'm only a science student."
"That's all the more reason to know!"

Materials
* *hard-boiled egg*
* *fresh egg*

To Do

Place a fresh and hard-boiled egg side by side. Then challenge a friend to identify the fresh and hard-boiled eggs by observing their spins.

While on its side, give one egg a whirl. Gently stop it in the middle of its spin. Release the egg. What happens?

Repeat this action with the other egg. How do these eggs differ in their behavior?

The Science

Inertia, inertia, inertia. It's that word again! To understand how inertia puts the spin on things, we must first examine the contents of these eggs.

The fresh egg has a liquid interior. While it spins, both its outer shell and liquid interior moves. When the egg is stopped, only the exterior comes to a halt. The liquid on the inside keeps spinning. So if the stopped egg is released, the spinning yoke gets the egg turning again.

The hard-boiled egg is solid on the inside. When it spins, both shell and hard insides turn. When it stops, everything stops completely. If the stopped egg is released, it will not start to spin again.

CHECK IT OUT! How could you use inertia to test whether there is ice or water inside a container?

1.6 GOOD CATCH!

Magicians, jugglers, and other performers often have a good understanding of physical science. Some know how to use optics to make an object disappear. Others understand acceleration and use it to juggle clubs. Still others are familiar with heat transfer and apply this knowledge to working with fire.

Congratulations. It's your turn to perform. For this act, you won't need a rabbit, sword, or burning torch. All you need is a stack of coins.

Materials
* *a bunch of coins*

To Do
Balance a stack of coins on your elbow. Snap your arm forward and catch the coins in midair!

The Science
Like many tricks, this one is not as difficult as it appears. That is because you have a helper: inertia. As you snap your arm forward, the coins are left unsupported in midair. Since they have inertia, they do not move sideways or drop too quickly. Instead, their drop begins slowly. By the time your wrist comes around, the coins have not had enough time to gain much speed. Instead, their relatively slow movement makes them an ideal target for a midair snatch!

CHECK IT OUT! Go watch a magician or street performer. Can you identify which tricks depend upon inertia's hidden help?

1.7 MARBLES IN A MUG

*N*ow that you've learned how to catch a stack of coins, how about catching marbles in a mug? As you may discover, this trick can get frustrating, especially if you don't uncover the secret of the technique. But here's a hint: This secret has to do with—you guessed it—the "I" word. Good luck!

Materials
• *several marbles*
• *mug*

To Do
Hold the mug's handle with your pinkie, ring, and middle finger. This leaves your index finger and thumb free. Using only these two fingers, pick up a marble. With an upward arm motion, toss the marble into the air (but keep holding the mug). Quickly position the mug beneath the falling marble. Great catch!

Now, it gets tougher. While grasping the mug with the marble still inside, pick up a second marble with your index finger and thumb. Now try throwing this second marble, so that it joins the first in the mug.

N O T E : If you were lucky enough to get the second marble in (using this same technique), try tossing and catching a third one.

The Science
The first marble is the easy one to catch. As it rises, you position the mug directly beneath it. Gravity does the rest.

The next marble is more difficult to snare (while keeping the first one within the mug). If you try the same technique, the captured marble keeps jumping out!

The best way to capture the second marble is using a "drop-n-slide" technique. Don't toss the second marble upward. Instead, release your grip on this marble. As it falls, quickly position the mug in its downward path. As you'll discover, the mug only needs to drop about an inch for the marble to clear the rim.

CHECK IT OUT! Take a bathroom scale into an elevator. Stand on the scale and ride up and down the floors (and try not to worry what the other passengers think). What happens to your weight as the elevator begins to go up? What happens to your weight as the elevator starts going down?

1.8 STACK SHOT

t's multiple guess time! Imagine a stack of four nickels set on a flat tabletop. Suppose you were to shoot a fifth nickel at the stack so that it hits the bottom coin dead center. What do you think will happen to the stack?

a. the fifth nickel will bounce off the stack

b. the fifth nickel will replace the bottom nickel

c. the fifth nickel will stop in front of the stack, but the bottom nickel will shoot out

And the answer is . . .

Materials
* five nickels

To Do
Stack four nickels on a slick, flat surface. Place a fifth nickel several inches away. Flick the nickel at the stack so that it strikes the stack squarely in the middle of the bottom coin. The flip doesn't have to be very powerful, just dead on center.

The Science of Off-Center Shots
The stack crumbles as this adventure goes down the drain. Glancing blows don't work.

The Science of Dead Center Shots
The energy of the collision is transferred to the coin at the bottom of the stack. As it moves off, the three coins above it fall straight down because of their inertia. The flicked coin transferred all of its moving (or *kinetic*) energy upon the collision to the bottom coin but it lacked the "extra" power needed to take its place beneath the remaining stacked coins. Therefore, the bottom coin flew off, but the fifth coin remained in front of the now three-coin stack.

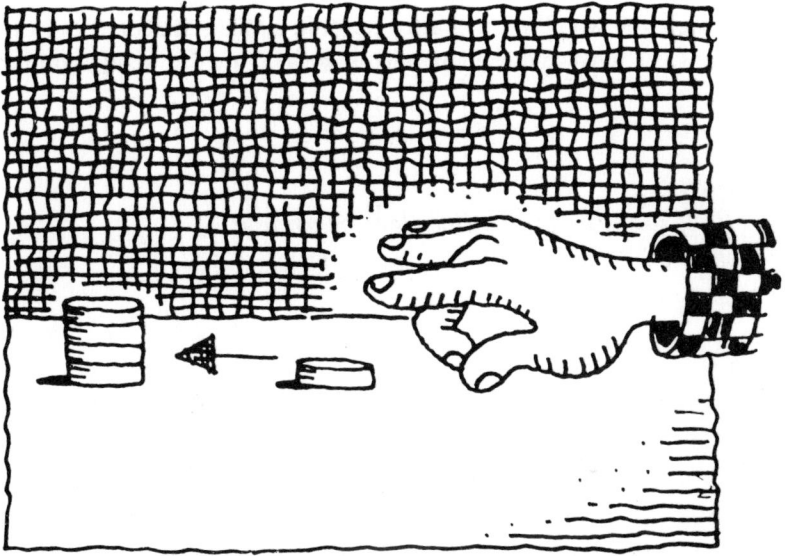

1.9 UPSIDE-DOWN WATER

May 2013

Have you ever seen a science fiction movie that takes place in a spinning space station? If so, chances are that the astronauts weren't weightless. Instead, the rotation produced some sort of artificial gravity that kept the astronauts against the station's edge.

Although you won't get a chance to build a space station (not in this book anyway), here's your chance to observe a down-to-earth spinning effect.

Materials
* small plastic sand bucket (toddler's beach toy)
* 2-foot long kite string
* a cup of water

To Do
Find a toddler who won't miss a sand bucket. Cut a section of kite string about 2 feet long. Tie the string to the arch of the handle.

Add about a cup of water to the bucket . Test the knot to make sure that it won't slip out.

Go some place outdoors or to an area where spilling water won't be a problem. Make sure you have plenty room around you so that you won't hit anyone or anything.

Now grasp the free end of the string and lift the bucket. Start swinging the bucket in a circle using a steady and gentle movement. Increase the force and angle of the swing so that the container "climbs" higher and higher. Keep going until the bucket loops completely upside down. What happens to the water?

The Science
The spinning motion produced a "force-like" effect that kept the water within the bucket. Since the effect of the spin was stronger than the attraction of gravity, even when the bucket spun through an upside down position, the water didn't spill out.

CHECK IT OUT! Take a ride on a spinning amusement park ride that uses this same effect. As you'll experience, these gravity spinners keep you plastered to the walls.

Centripetal.

Balloon + penny.

Watch youtube.

1.10 WITH OR WITHOUT INERTIA

✓ May 2013

ere's another one of those adventures that you can present with all of the mystery of magic. Tell your audience that you can snap the thread either above or below the fishing sinker. Let them decide on where they want you to break the thread. Then snap the string according to their wishes.

Materials
* *2 ounce fishing sinker*
* *fine sewing thread*

To Do
Break off two pieces of thread, each about 6 inches long. Tie both threads to the "eye" of a fishing sinker. Tie the free end of one thread to a secure support so that the sinker hangs freely.

Grab hold of the other free end of the hanging thread. Gradually increase the force of your pull. What happens? Where does the thread break?

Replace the broken thread. Instead of a gradual pull, this time give the lower thread a quick snap. Where does the thread break? Can you explain your observations?

The Science
When you pulled down slowly, you overcame the inertia of the sinker. In fact, the sinker's weight added to your pulling force to place a greater tension in the thread above the sinker than in the thread below the sinker. This caused the thread to snap above the weight.

When you snapped the thread quickly, the inertia of the sinker worked against the pull. Its resistance opposed the pull. This created a greater stress beneath the sinker, causing the break to occur in this lower segment.

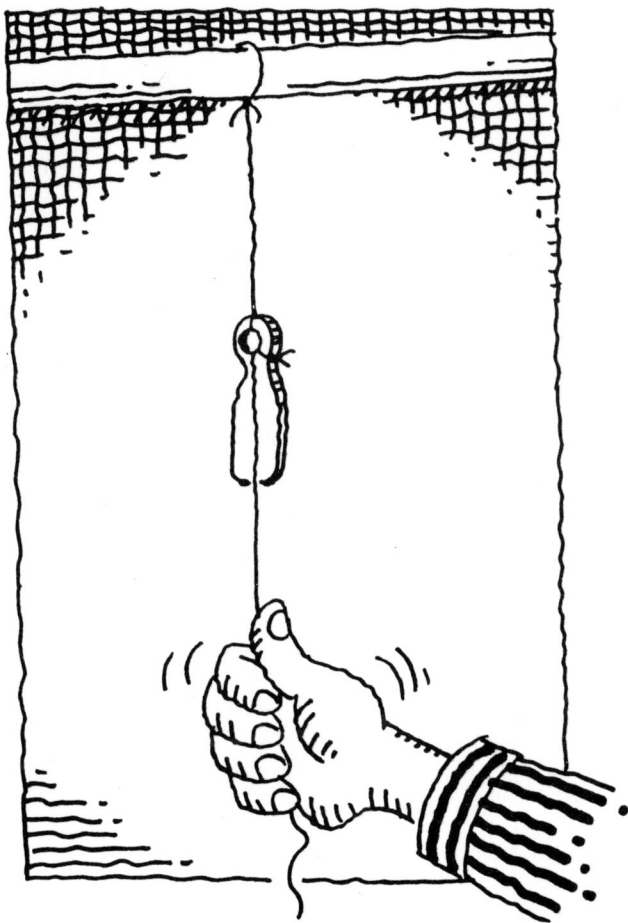

IN THE BALANCE

2.1 CENTER OF GRAVITY

Every object has a special point called its *center of gravity*. If you support an object so that it is allowed to spin freely, it will eventually come to rest so that its center of gravity is directly below the point of support. If that's a bit unclear, this experiment will hopefully set things straight.

Materials

- *heavy stock paper or cardboard*
- *6-inch-long kite string*
- *pair of scissors*
- *pushpin*
- *marker*
- *heavy stock paper*
- *washer*

To Do

Trace or copy the picture on the next page onto a sheet of heavy stock paper or cardboard. Cut out the drawing.

Tie a washer to a 6-inch long kite string. Press a pushpin through the circle marked A into a wooden support. The pin shouldn't be snug. It should be loose enough to allow the rocket to freely rotate.

Tie the free end of the string to the pushpin shaft. Use a marker to trace the path that the string makes. Repeat this step by tracing the lines made when the pin is placed into points B and C. Do the three lines intersect? What's so special about this point?

The Science

Everything tends toward stability. The stability (balance) of the rocket was affected by the irregular distribution of weight. When the pin was placed into point A, the rocket was free to rotate into its most stable position. The drawing came to rest so that its center of gravity was directly below the pin (point of support).

When the pin was placed at the two other points, the same thing happened. The rocket came to rest so that its center of balance was directly beneath the pin. The three lines intersected at the rocket's center of gravity.

CHECK IT OUT! Can you cut out a flat object whose center of gravity isn't located on the object?

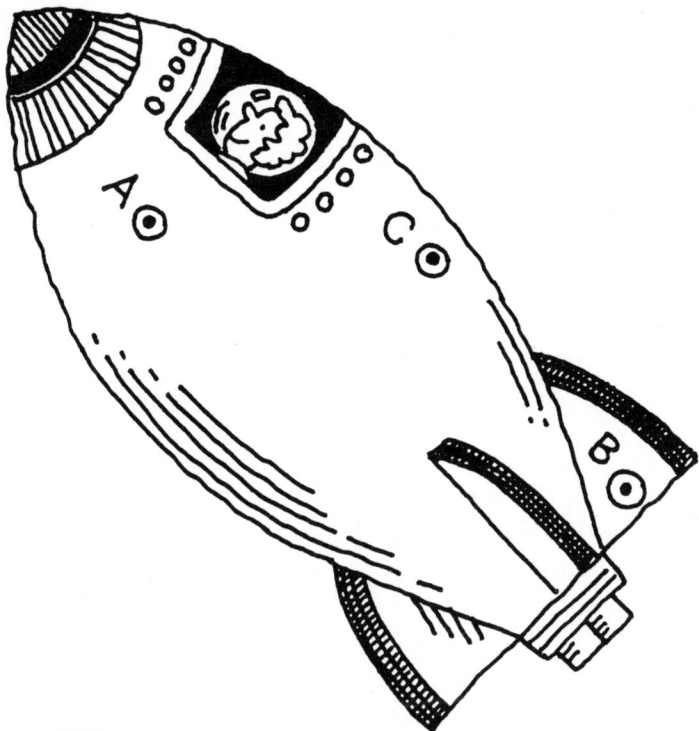

2.2 BALANCING SPUD

*L*adies and gentleman. Please direct your attention to the rim of the plastic drinking glass. You are about to witness an incredible feat of balance, poise, and talent. For soon you will observe a mindless spud balancing upon the edge. And there it will remain, even as the cup of water upon which it rests, is raised and emptied.

Materials
- *two forks*
- *small potato*
- *toothpick*
- *two large cups (one filled with water)*

To Do
Insert the forks into the potato as shown. Then insert a toothpick into the bottom of the potato. Position the toothpick on the rim of a water-filled cup. If the potato doesn't balance, adjust the forks accordingly. Steadily raise the water-filled cup (with balancing spud) and pour its liquid contents into an empty cup.

The Science
The forks add stability to the potato by shifting its center of gravity. As you learned in the previous adventure, the center of gravity is critical to an object's balance.

Unlike the rocket drawing, the potato has bulk. Its point of balance is located in 3D space. When the forks were added to the potato, a new center of gravity was created. This center was located directly beneath the toothpick. When the potato was placed on the rim, it moved to a balance point where its center of gravity was directly beneath the spot where the toothpick and rim contacted.

CHECK IT OUT! Construct a "high wire" setup using kite string. Test your spud with this new challenge.

2.3 SLIP SLIDER

Have you ever heard of places where cars appear to roll uphill? Weird, huh? Well, this next experiment will have funnels going uphill! These funnels, however, aren't hypnotized with anti-gravity magic. Their motion is produced by a constantly changing center of gravity!

Materials
* *two identical funnels*
* *tape*
* *four books*
* *paper clip*
* *two long dowels (at least 1 foot long)*

To Do
Place the mouths of two funnels together and tape their joined rims.

Now tape the ends of the dowels as shown in the illustration. The taped ends should have some flexibility to move.

Make a stack of three books. Position the untaped ends of the dowels atop the stack. Position the taped ends of the dowels on a single book.

Place the joined funnels between the dowels at the lower end stack. Slowly open the gap by bringing the dowels apart at the upper stack. As the gap widens, the funnels will climb up the dowels. The object of this game is get the funnel stack to climb to the top of the dowels. Think you can do it? Give it a try.

The Science
Although the funnels appear to be going against gravity, they're not. As the gap widens, the funnels take on a new center of gravity. To stabilize, the funnels drop slightly and spin forward. The forward part of the motion makes it appear as if the funnels are climbing upward. In actuality, they're slipping downward.

CHECK IT OUT! Build an anti-gravity game. Replace the funnels with a ball and use longer dowels. You may want to build a more stable base using wood or foam core. The object of the game is to bring the ball to the top of the gap.

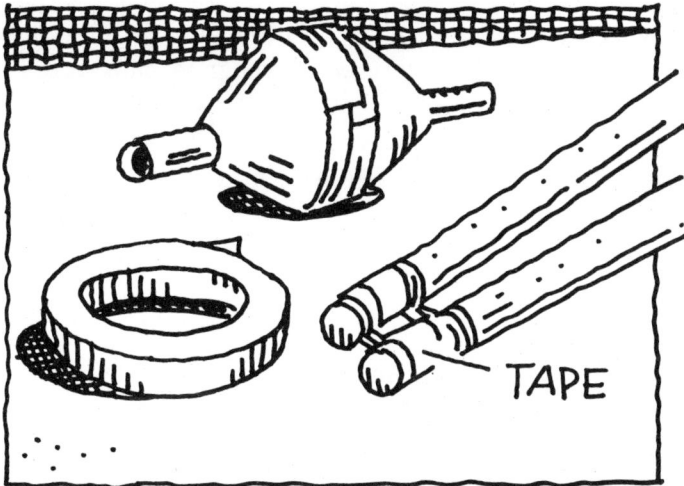

TAPE

2.4 FLOATING FLAVOR

ow can you tell a regular soft drink from its diet form? Most likely, you can taste the difference. Although manufacturers of artificial sweeteners like to believe that their "sweetening" chemicals taste just like the real thing, they don't. Most people can detect a chemical aftertaste. There's another difference too. This one, however, is seen, not tasted.

Materials
- a can of regular (sweetened with sugar) soft drink
- a can of diet soft drink
- fish tank or large bowl filled with water

To Do
Place a can of regular soft drink in a fish tank or large bowl filled with water. Does it float or sink? Now place a can of diet soft drink in the bowl. What happens to this can?

The Science
Floating is all about the balance of forces. The downward force that causes something to sink is called *weight*. The weight depends upon the amount of matter "stuffed" in an object. The upward force that causes something to rise is called *buoyant force* (or B.F.). The B.F. depends upon the amount of space the object takes up.

When the weight is greater than the B.F., the object sinks. When the B.F. is greater than the weight, the object floats. When the weight and B.F. are equal, the object remains at the same level (neutral buoyancy).

Regular soft drink is a solution of carbonated water, flavoring, and plenty of sugar. Since the soft drink's sugar is dissolved, you can't see it. You can, however, certainly taste it.

Like regular soft drink, the diet form contains carbonated water and flavoring. The sugar, however, is replaced by a small amount of a artificial sweetener. The taste of the substitute is so strong that very little is needed.

Cans of regular and diet soft drink are almost neutral in buoyancy. The extra sugar in the regular soft drink, however, makes this can slightly heavier. The diet soft drink, although it has a small amount of artificial sweetener, lacks the sugar and is slightly lighter. Although the difference is small, it is significant enough to cause the can filled with the regular soft drink to sink, while the can filled with the diet soft drink floats.

CHECK IT OUT! Design a way to find out how much sugar (in grams) is added as a sweetener to regular soft drink.

2.5 SALT FLOAT

"**N**O SINKING ALLOWED!" The water that fills Salt Lake contains a great deal of dissolved salt. This salt makes the water "extra" buoyant. It is so buoyant that it's nearly impossible for a person to sink! But you don't have to go to Salt Lake to observe extra buoyancy. You can produce this effect right in your kitchen. Here's how.

Materials

* raw egg
* spoon
* ½ cup salt
* tap water
* two tall plastic containers
* measuring cup

To Do

Fill two tall plastic containers ¾ full with water. Add about ½ cup of salt to one of the containers. Mix thoroughly until the salt dissolves and then add more salt. Keep adding salt until no more salt can be dissolved. Don't worry about adding too much!

Carefully place a raw egg into the container filled with pure tap water. What happens to the egg? Now remove the egg and place it into the container filled with salt water. What happens to the egg this time?

The Science

As you've learned, floating is the delicate balance between weight and buoyant force (B.F.). In this activity, you didn't change anything about the object. Instead, you changed the surrounding environment.

The strength of a B.F. is dependent upon how much "stuff" is dissolved in the solution. The more dissolved materials, the greater the B.F. When you added salt to the tap water, you not only made the solution saltier, but you increased the amount of B.F. it could produce.

When the raw egg was placed in tap water, it sunk. The egg's weight was slightly greater than the B.F. it generated. When this same egg was placed in salt water, it floated. Although the egg's weight stayed the same, the buoyant force was greater. This greater B.F. offset the egg's weight and prevented it from sinking.

CHECK IT OUT! Design an experiment that would compare the buoyancy of a sugar-and-water solution with a solution of salt water.

2.6 GOING UP?

A hydrometer is a tool that measures the concentration of a solution. It is simple and easy to make. It's even simpler to use. Here's a hydrometer that you can use to measure the relative concentration of all sorts of solutions.

Materials
- drinking straw
- pair of scissors
- marker
- ruler
- waterproof clay
- three tall containers
- salt
- measuring cup

To Do
To make the hydrometer, cut a section of drinking straw that is about 4 inches long. Use a ruler to place ¼-inch marks along the length of the segment. Place a small lump of waterproof clay (about the size of a pea) in one end of the straw.

Fill all three containers ¾ full with water. Add about ¼ cup of salt into one container. Add about ½ cup of salt to another container. Mix both containers thoroughly. Make sure that all the salt dissolves.

Place the hydrometer into the pure water solution. Observe and record the mark that the water level makes with the hydrometer scale. Remove and rinse the hydrometer in tap water.

Place it into the solution that contains ¼ cup of salt. Again, observe and record the mark of the water level. Remove and rinse the hydrometer in tap water.

Place the hydrometer into the solution that contains ½ cup of salt. Observe and record the mark of the water level.

The Science
This experiment, like the previous egg float, explores the relationship between dissolved material and buoyant effect. The more material (in this case, salt) that is dissolved in water, the greater the solution's B.F.

The increase in buoyancy is measured by the height at which the hydrometer floats. A large amount of dissolved salt produces a strong buoyant effect. The "extra" lift pushes more of the hydrometer out of the water. In contrast, tap water produces the least lift.

CHECK IT OUT! The battery in your car contains a dangerous acid. Its concentration is measured by mechanics using an acid-resistant hydrometer. Their garage tool works just like the one you just constructed. When placed in battery acid, it floats at a height which indicates the acid's concentration.

2.7 DESTINY IN DENSITY

Well, it finally had to happen. So far, we've talked about floating without mentioning the "D" word, *density*. We've explained floating in terms of a balance between buoyant force and weight. There is, however, another way of presenting buoyancy. It has to do with a concept called *density*.

Density refers to the concentration of matter. Solutions that have a great deal of dissolved materials are denser than those that have nothing dissolved in them. For example, salt water is denser than plain old tap water. Likewise, a one-quart solution with ½ cup of dissolved salt is denser than a similar solution with only ¼ cup of dissolved salt.

Now that you understand density, here's how it fits in with floating. When two substances are mixed together, the denser material sinks. Chocolate syrup sinks in milk. Balsa wood floats on water. Carbon dioxide shot from a fire extinguisher sinks in air.

Materials

* *tall drinking glass*
* *cup*
* *corn syrup*
* *cooking oil*
* *water*
* *food coloring*
* *balsa wood*
* *lid to a plastic film canister*
* *piece of hard sugar candy*
* *penny*

To Do

Carefully pour several ounces of corn syrup into a tall drinking glass so that the corn syrup forms a layer several fingers thick. Pour about 8 ounces of water into a small, clean cup. Next, add several drops of food coloring to the water (it makes the layers look more dramatic). Tilting the glass, carefully pour this colored water into the drinking glass so that the water flows down the inner side of the container (and doesn't splatter). Save a small amount of the colored water for later. After the water has been added, introduce the cooking oil in the same manner. Stop when the oil forms a layer that is several fingers thick.

Observe how the layers interact. Do they mix? Do they stay perfectly separated? Now add a piece of balsa wood, the lid to a plastic film canister, a piece of hard sugar candy, and a penny to the mixture. At which level do these materials float?

The Science

The mixture that you just formed has three layers. The layers remain separated due to their difference in density. Eventually some mixing (mostly between the corn syrup and dyed water) will occur. Until then, this remains a three layered mixture. Since the mixing doesn't occur on an atomic level, this mess is NOT a solution.

The layers stay separated because each has its own density. The corn syrup has the greatest density and remains at the bottom. The colored water has a medium density and remains in the middle. The oil is least dense and remains afloat.

Balsa wood has a density less than oil. Therefore, the wood scraps floated on top of the oil. The plastic from a film canister is less dense than water, but more dense than oil. You can tell that by observing the layer at which it floats. Hard sugar candy is denser than water, but less dense than corn syrup. The penny is the densest one of them all. It sinks to the bottom of the container.

By now, you may be saying to yourself, "Hey, wait a second. This experiment is rigged! The corn syrup stayed on the bottom, because it was poured in that layer. Likewise for the water and oil."

You are partly right. Yes, the liquids were poured in order to help produce distinct layers. Their position, however, was determined by their density. To prove this, repeat the mix but this time start with the oil first.

CHECK IT OUT! Which is denser, hot air or cool air?

2.8 UNDERCURRENTS

Powerful currents exist beneath the ocean's surface. Unlike the surface variety, these unseen currents are not created by the wind. Instead, their movement is driven by differences in density. In fact, these subsurface currents are called *density currents*.

When salt water evaporates or becomes trapped as ice, its load of dissolved salts remains behind. This "leftover" salt dissolves in the remaining water to make it denser. If unmixed, this dense water sinks through the less saline water below. As it sinks, the water spreads out, forming the sideways movement of a density current.

Materials
- *salt*
- *blue and red food coloring*
- *ice*
- *spoon*
- *two cups*
- *two clear plastic pitchers*

To Do
Fill a cup halfway with ice water. Add several drops of blue food coloring.

Fill a second cup halfway with tap water. Dissolve as much salt as possible into it. Add several drops of red food coloring to this saline water.

Fill a large pitcher halfway with room temperature water. Slowly pour the dyed ice water into this pitcher. Tilt the pitcher so that the ice water runs down its inner side. Observe what happens when the ice water meets the room temperature water. Does the ice water float or sink? Is ice water more dense or less dense than room temperature water?

Fill a second pitcher halfway with room temperature water. Slowly pour the dyed salt water into this pitcher. Tilt the pitcher so that the salt water runs down the inner side of the pitcher. Observe what happens when the salt water meets the tap water. Does salt water float or sink? Is salt water more dense or less dense than tap water?

The Science

When water is cooled, its tiny particles (molecules) slow down. Their slower speed allows the molecules to pack closer together. The result is an increase in density. Therefore, when cold water is poured into room temperature water, it sinks to the bottom of the container.

In the "Going Up" experiment (see p. 40), you've learned that salt water is denser than tap water. Therefore, it should make sense that salt water "pours" through tap water and forms a separate lower layer.

CHECK IT OUT! Reverse the experiment. Add tap water to a pitcher that is half filled with ice water. Then add tap water to a pitcher that is half filled with salt water.

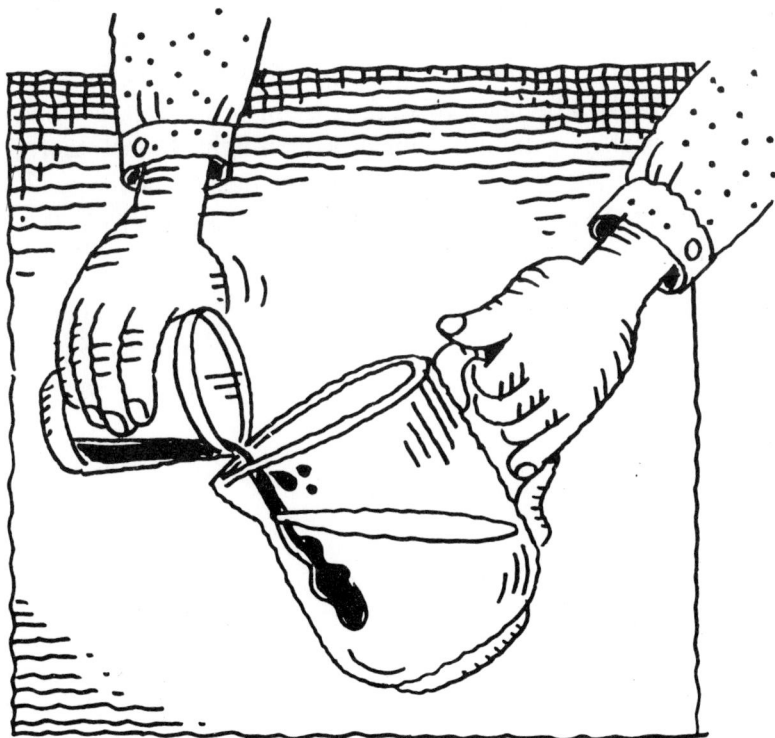

2.9 TAKING A DIVE

Where would you expect to find a rock that floats? In a fantasy story? In a science fiction movie? Try a drug store. Strange, but true: Floating rocks are sold at most drug stores!

Materials
* pumice (available at drug or cosmetic stores)
* small container filled with water

To Do
Take a piece of pumice and place it in a small container filled with water. Does the rock sink or float?

The Science
Pumice is a volcanic material that looks and feels like hardened foam. It is formed when molten rock, which was filled with boiling gas, cools very quickly. As it becomes solid, the rock material traps gas bubbles.

The air pockets within the pumice keep it light. In fact, there is so much trapped air that if you averaged the density of the solid rock material and the air spaces, you'd have an overall density that is less than water. Hence, pumice floats!

2.10 PUMICE IMPOSTORS

Other materials besides pumice can trap buoyant air bubbles. Even a lowly piece of aluminum foil can trap enough air to stay afloat. Here's how:

Materials
* small square of aluminum foil
* small container
* water

To Do
Crumple a small square of aluminum foil into a loose ball. The ball should be about the size of a large marble. Make sure there are many nooks and crannies within the material in which air bubbles can fit.

Drop the crumbled foil ball into a container of water. The ball should float on the surface of the water. Now shake the aluminum ball beneath the surface of the water. Keep shaking it until there are no visible air bubbles on the foil's surface. Release the ball and this time the ball should sink.

The Science
The overall density of the foil ball determines its buoyancy. When it was first crumpled, the ball contained many air spaces and bubble-trapping nooks and crannies. These extra air spaces decreased the overall density of the ball and made it less dense than water. As a result, the ball floated.

When the bubbles were shaken away, the overall density of the ball increased. With this greater density, the foil ball sank.

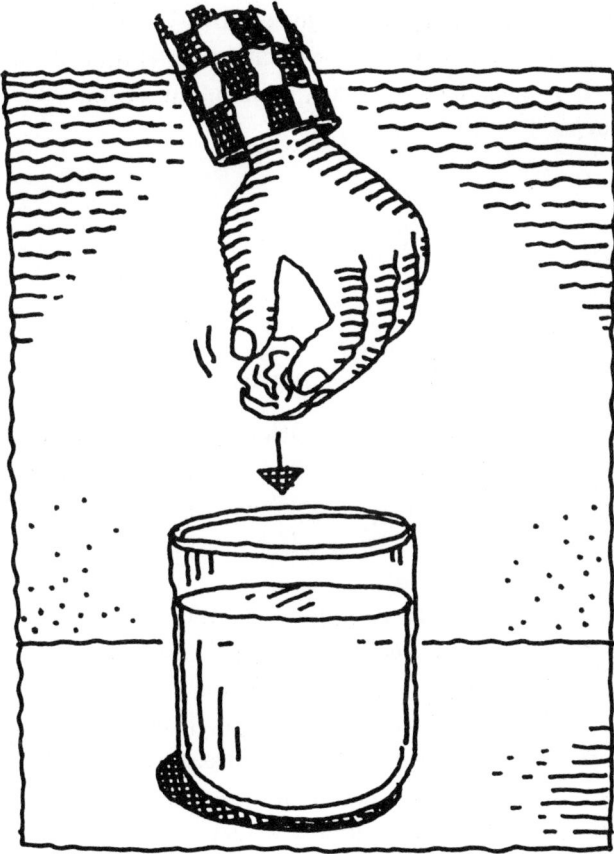

2.11 UP 'N' DOWN SPAGHETTI

hen something goes down, does it always come up? Well, not always—especially when it relates to sinking. However, here's an experiment that produces a steady column of rising and sinking spaghetti pieces.

Materials
- a strand of uncooked spaghetti
- a can or bottle of club soda
- tall, clear plastic cup

To Do
Fill a tall, clear plastic cup halfway with club soda. Break into 1-inch pieces a strand of uncooked spaghetti. Put the spaghetti pieces into the cup. Observe their behavior for the next 10 minutes.

The Science
Club soda is a solution which contains carbon dioxide gas that is dissolved in water. While dissolved, the carbon dioxide molecules are separated, dispersed, and mixed among the water molecules.

When the club soda container is uncapped, pressure is released from the surface of the soda. The lowered pressure allows the carbon dioxide gas to exit the solution and form tiny gas bubbles.

When the carbon dioxide bubbles attach to the surface of the spaghetti, they provide "extra lift." As attached bubbles enlarge, the overall density of the spaghetti pieces decreases. Eventually, enough bubbles lift the pieces to the surface.

When the pasta reaches the surface, things change. Some of the bubbles burst. Other bubbles become detached. As the bubbles disappear, the spaghetti's overall density increases. Soon, it becomes denser than the surrounding water and sinks.

CHECK IT OUT! Ever hear of the "bends?" It's a painful disorder that affects scuba divers when they surface from the ocean too quickly. When a diver is beneath the ocean surface, he or she breathes pressurized air. This air, like that at the surface, is made up mostly of nitrogen and oxygen.

During their dive, the inhaled nitrogen dissolves into the diver's blood. As the diver ascends, the surrounding pressure drops. This allows the nitrogen to come out (like carbon dioxide in the unsealed club soda). As the nitrogen exits the diver's blood, it forms tiny bubbles. These bubbles collect near the diver's skeletal joints to cause painful and sometimes fatal symptoms. Often the victim doubles over in pain. Hence, that's how the "bends" got its name!

2.12 CARTESIAN DIVER

A pump spray perfume bottle and a Cartesian diver are two very different items, yet they share some basic physics. When you push down on the pump, you produce a force. This force pushes on the perfume stored within the bottle. The pressurized liquid rises through a tube and emerges as a mist through the spray nozzle.

But what has squeezing a pump spray bottle have to do with a submarine-type toy that dives on command? Read on.

Materials
- *glass medicine dropper with bulb*
- *1- or 2-liter plastic soft-drink container with cap*
- *water*

To Do
Fill the soft drink container to the brim with water. Draw water into the dropper until the glass pipette is about ⅓ filled with water. Put the dropper into the container. If needed, add more water to the container. Tightly screw on the cap.

Grasp the container. Slightly increase the pressure of your grip. Observe the appearance of the trapped air bubble within the dropper. Keep squeezing and you'll soon discover why this device was given the name diver.

The Science
When you squeezed the container, you produced a force. This force acted upon the water in the container. Since the water could not be easily compressed, the force was transmitted "through" the water. Within the diver, the force pressed upon the air bubble. Air, unlike water, is easily compressed. Under the stress of this force, the bubble shrank in size. As it shrank, its buoyant force (B.F.) decreased. As soon as the B.F. became less than its weight, the diver sank.

CHECK IT OUT! Build a Cartesian diver out of aluminum foil. Make sure that your design not only rises and sinks at command, but also remains stable within the water container.

2.13 BLOW ALL BALLAST!

A submarine's ability to sink and surface is set by the balance of forces between the sub's weight and its buoyancy. Although the weight of the hull remains mostly the same, the buoyant force is increased or reduced.

Submarines have compartments called ballast tanks. Ballast tanks can be opened and closed. In order to dive, the tanks open. As they fill with water, the sub becomes less buoyant and sinks. In order to surface, pressurized air is injected into the tanks. This air blast "blows out" water to make the submarine more buoyant.

Materials

* film canister with lid
* string
* tape
* ¼ teaspoon baking powder
* spoon
* 1 ounce sinker (fishing weight)
* manual hand drill
* large bowl
* water

CAUTION

Handle baking powder with care. It can be harmful if swallowed, inhaled, or rubbed in the eyes. Also, have an adult drill the holes in the canister.

To Do

Have an adult use a hand drill to carefully place six small holes (each about ⅛ inch in diameter) in the bottom of a film canister.

Tie a string to a 1-ounce fishing sinker. Use tape to secure the sinker's string to the bottom of the film canister. Place about ¼ teaspoon of baking powder into the canister. Seal the canister with its lid.

Place the canister and sinker in a large bowl of water. As the canister fills with water, it becomes less buoyant. When it begins to sink, release the canister. Keep watching. In a few moments, the evolving gas should turn things around.

The Science

When the canister first filled with water, it sunk. Within moments, however, the water that entered the canister mixed with the baking powder to produce carbon dioxide gas. As the gas was generated, it rose to the top of the canister. There the gas bubbles took the place of water. Water that had previously occupied this space was forced downward out of the canister.

This action of replacing one substance by another is called *displacement*. When enough water had been displaced, the B.F. was strong enough to bring the canister to the surface.

BAKING POWDER

2.14 CONCRETE SHIPS?

Ship hulls made of concrete? Although this may seem crazy, it's not. There are thousands of ships with hulls made of concrete. In fact, using concrete as a hull material should seem no more ridiculous than building hulls out of steel.

Have you ever wondered why ships made of a material heavier than water can float? What keeps them from sinking to the bottom?

Materials
* *waterproof modeling clay or soft clay used by florists*
* *several pennies*
* *a small basin*
* *water*

To Do
Roll up a ball of clay. Gently place it on the water's surface. Release it. What happens?

Now shape the lump of clay into a small boat. For a real challenge, you'll want to build a clay vessel that can support the weight of several pennies.

The Science
So far, floating has been explained in terms of two concepts: 1) the balance between weight and buoyant force; 2) the differences in density.

Here's a third concept that can be applied to floating: displacement.

When you sit in a bath tub, the level of water rises. It increases because your body now occupies a space that was previously filled with water. When you sat down, you pushed away (or displaced) an equal volume of water. When you displace water, you generate an upward force that is equal to the weight of the water you "pushed out."

A ship's hull also displaces water. It is designed to displace enough water to keep the entire boat (hull, decks, stacks, etc.) afloat.

2.15 BALLOON ELEVATOR

n the Jules Verne story, *The Mysterious Island*, several people are trapped on an isolated island. To escape, they raise a sunken wreck by stuffing a large hot air balloon into its hull and pumping it full of air. As the balloon inflates, it displaces water that had filled the sunken hull. When enough water is displaced, the ship rises to the surface.

Materials
* paper cup
* 1 ounce lead sinker
* string
* flexible straw
* water-filled container
* pencil

To Do
Using a pencil, poke two small holes beneath the rim of the paper cup. The holes should be positioned at opposite sides of the cup.

Cut a length of string about 8 inches long. Insert the string into one of the holes and tie it. Then pass the string through the "eye" of a sinker and through the cup's other hole. Fasten the string to the cup.

Place the cup and sinker into a container filled with water. Tip the cup so that it fills with water. The weight of the sinker should bring the cup to the bottom of the container.

Make a small bend in the flexible collar of the straw. Insert this bend into the cup. Place the other end of the straw in your mouth and blow. What happens to the cup?

The Science
As you blew, air traveled along the straw. When it exited the straw, it floated upwards as a stream of bubbles. These bubbles soon became trapped within the inverted cup. The air that rose to the top of the cup displaced the water. Eventually, enough air filled the cup to produce a strong lifting force.

CHECK IT OUT! Shipyards around the world have large floating "garages" called dry docks. A dry dock looks like a large bathtub with a hinged door. Ships sail into the dry dock's confined space. The dry dock is sealed and air is pumped beneath the floor of the dock. As air displaces water, the floor becomes more buoyant. This rising floor can gain enough lift to raise an ocean liner out of the water!

WHAT'S THE ATTRACTION?

3.1 PENNY POOL

ave you ever examined a drop of water that had fallen onto a sheet of wax paper? If so, you know that the drop wasn't flat. Instead, it formed a bead. This round shape results from tiny, unseen forces that exist in the smallest particles of water.

Materials
* penny
* quarter
* water dropper
* water

To Do
Make a guess. How many drops of water can fit on the head of a penny without spilling over the edge? Write down your guess.

Now use a dropper to carefully add water to the head of a penny. Count each drop. Try not to splash the drops or move the penny. When you have finished, compare the actual number of drops with your guess. How close were you?

Let's make another guess. Based upon what you have observed, predict the number of drops that can fit on the head of a quarter. Then check your answer.

The Science
Water is made up of very tiny particles called water molecules. Each water molecule is made up of two hydrogen atoms and one oxygen atom (hence, its formula H_2O). This molecule can be represented by something that looks like this:

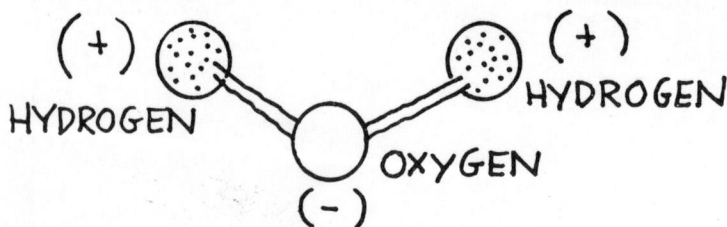

$(+)$ HYDROGEN $(+)$ HYDROGEN   OXYGEN $(-)$

But, there's more. Each hydrogen end takes on a positive charge. Each oxygen takes on a negative charge. The charges influence the way each molecule lines up with its neighbor. The negative end of one molecule lines up with the positive end of its neighbor. The positive end of another lines up with the negative end of its neighbor. This attraction results in a type of "skin" that holds together each bead of water. The skin is strong enough to stretch and hold in additional water molecules. That is why so many drops fit on the head of the coin. Break the attraction between neighboring molecules and the bead flattens by spilling water over the coin's rim.

CHECK IT OUT! Compare the way water "beads" behave on a waxed and unwaxed wood surface.

3.2 FILLED TO THE BRIM?

*S*tep right up and make your guess!

Now that you understand water tension it's time to play "Quarter in the Cup."

To win, all you have to do is guess the numbers of quarters that can be placed into this cup without water spilling over the rim.

Materials
* clear plastic cup or glass
* stack of quarters
* water

To Do

Fill the cup or glass to the brim with water. Examine the cup. Guess how many quarters can be placed into it without spilling water over the rim. Write this guess down. Now see how close you are to your prediction by gently dropping quarters into the cup.

The Science

Like the previous experiment, this one is based upon the attraction between neighboring water molecules. As quarters are lowered into the cup, they displace (or "push out") an equal volume of water. The water level rises.

Although the water rises above the rim, it doesn't spill over. It's that attraction between the opposite charges of oxygen and hydrogen that keeps the "skin" together. The skin, however, does have limits. When the surface becomes too curved, the balance is destroyed and water spills out over the rim.

3.3 SCIENCE ON A KITE STRING

"What a mess!"

If you spill something in the kitchen, it'll be a mess. If you spill something in the laboratory, it can be deadly!

Scientists use tricks to transfer liquids. These tricks prevent splashing and make sure that liquids pour into beakers instead of running down onto sneakers.

Materials
* water-filled pitcher
* 1-yard-long kite string
* sink
* cup
* a pair of scissors

To Do
Cut a length of kite string about a yard long. Now go to a sink and wet the string.

Tie one end to the handle of the pitcher. Run this string around the pitcher so that it keeps in contact with the pouring spout.

Position a cup in the middle of the sink. Have a friend hold the other end of the string over the center of the cup.

Make sure that the string is pulled tight. Then gently pour out the water. What happens?

HINT
Although you might be able to do this experiment on your own (using duct tape to secure the string), it's more fun to do it with a friend.

The Science
Water flowing over the spout runs down the string. As long as the string is keep taunt, water will flow along its length. At the bottom of the string, the water passes into the cup (without a mess).

Remember how water molecules beaded up on a sheet of wax paper? That was due to surface tension. Well, an attraction between molecules is at it again. This time, it keeps the water molecules attached to each other in a stream-like chain.

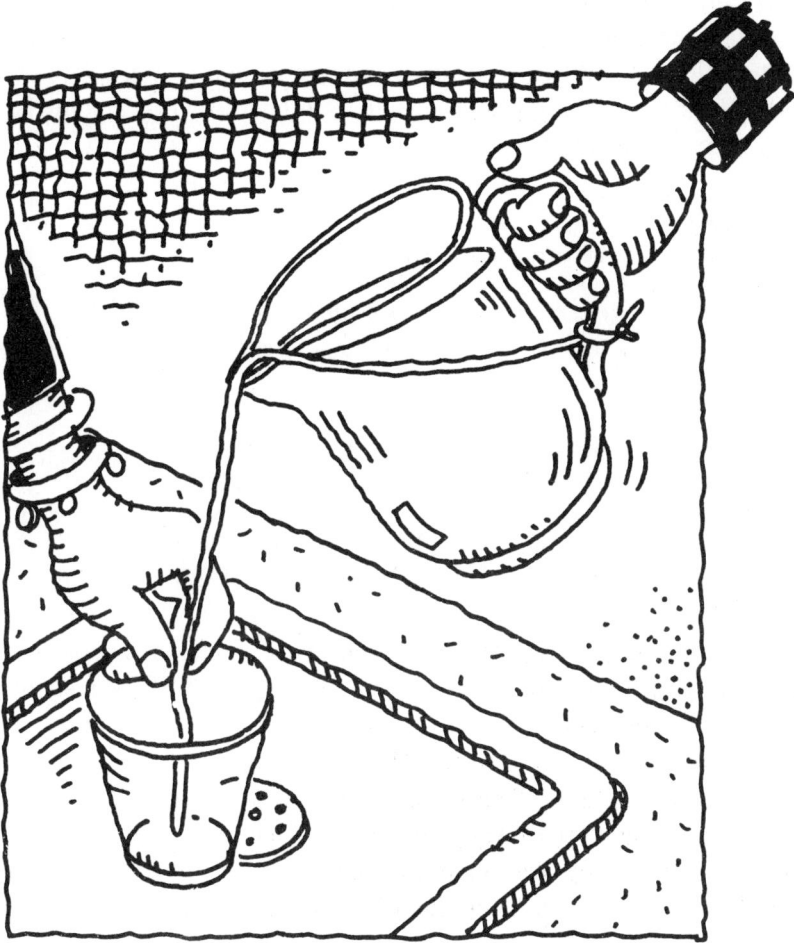

3.4 SIX TO ONE

*S*lowly to the right. Jump to the left. To the right again....just a little. Down quickly.

Water droplets appear to follow a choreographer's directions when flowing down a window. These directions seem to be laid out in a pattern of droplets. When a droplet encounters another droplet, they join. Quickly the two fall and encounter other droplets. Eventually, they drip out of sight!

Materials
* milk container or large drink box
* sharp pencil
* ruler
* tape
* water
* sink

To Do
Use a ruler to draw a horizontal line across the bottom edge of a milk container's side. The line should be drawn about 2 inches above the container's bottom.

Make six marks along the line, each about ½ inch apart. Use the pencil to punch out a hole at each one of the marks. Place a strip of tape across the holes.

Fill the container with water. While holding the container over a sink, remove the tape. Observe the six separate water streams.

Use your fingers to slowly pinch the jets together. What happens to the streams as they are being pinched? How long will they remain together? How can you make them flow separately again?

The Science
As you probably guessed, the streams' behavior is due to the attraction between water molecules. At first, the individual water streams are too far apart to affect each other's path. However, when they are pinched, the streams are brought together. Their force of attraction is strong enough to prevent the streams from separating. However, if a

hand physically separates them, they will go back to flowing as six individual streams.

CHECK IT OUT! Repeat this experiment with six holes placed vertically instead of horizontally.

✳ 3.5 WATER'S SKIN

ater striders are long-legged insects. Like other insects, they have three pairs of legs. Their legs, however, aren't used for walking on land. Instead, water striders walk—more like dart—on the surface of ponds, lakes, and streams.

This insect's body stands atop of the water's skin. As you are about to discover, this skin can support things other than insects.

Materials
* two paper clips
* water
* cup
* dishwashing liquid

To Do
Fill a clean cup with water. To build the delivery clip, bend apart the two big loops of a paper clip so that it looks like this:

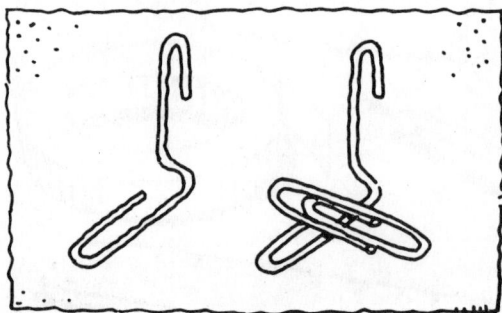

Hold this delivery clip so that one part is horizontal. Place the other paper clip on this flat part. Slowly lower the clip into the cup of water. Make sure that the clip is completely flat when it touches the water. As you keep lowering the delivery clip, the other clip should remain afloat on the water's surface.

Add a single drop of dishwashing liquid to the water. Observe the behavior of the floating clip as the soap spreads out.

The Science
The water's surface forms a skin that is tough enough to support a paper clip. (You might try testing the strength of this tension with

larger and heavier objects.) As long as water molecules stick to each other, the invisible skin holds together.

When you added detergent, soap molecules entered into the solution. They quickly dispersed throughout the water. Some of the soap molecules pushed apart neighboring water molecules. This broke up the attraction among the water molecules and caused the water tension to break up.

3.6 PEPPER PLATE

7-2-3 Pull! 1-2-3 Pull!

Did you realize that there is a miniature version of tug-of-war being played across the water's surface? One molecule tugs this way. Another molecule pulls the other way. But since the tugs occur equally in all directions, they cancel each other out.

A little bit of soap, however, can change the balance!

Materials

* *shallow plate*
* *water*
* *pepper*
* *dishwashing liquid*

To Do

Clean and rinse a large shallow plate. Make sure that all soap has been rinsed from its surface. Fill the plate ¾ full with cold water. Let the water stand until it is perfectly still.

Sprinkle some pepper across the surface of the plate. Add a single drop of dishwashing liquid near the rim of the plate. What happens?

The Science

The pepper was supported on a layer of surface tension. Within this layer, molecules of water pulled against each other. Since they pulled equally in all directions, the layer remained stationary.

The soap that was added to the water broke the surface tension. Since the forces were no longer active in this region of the plate, the surface tension on the far side of the plate caused the layer to contract. The pepper, riding atop of this layer, was carried across the surface.

CHECK IT OUT! Repeat this activity except, this time, substitute cooking oil for detergent. Does it work?

3.7 SPREADING LOOP

Have you ever wanted to choreograph a dance? Well, here's your chance. Your dancer, however, won't be made of flesh and bones. Instead, it will be made of cotton.

Materials
* 6-inch-long thread
* large pie pan
* dishwashing liquid
* water

To Do
Clean a large round pie pan. Make sure that all soap has been rinsed from its surface. Fill the pan about half full with cold water. Let the water stand until it is perfectly still.

Tie a 6-inch length of string into a loop. Bunch up the thread so that the loop has an irregular shape. Place this "squished" loop onto the surface of the water. Add a drop of dishwashing liquid to the center of the shape. What happens?

The Science
When dropped onto the surface, the thread floated atop a layer of surface tension. Adding soap to the water broke the surface tension. Since the surface tension was still active around the rim of the pan, this force "stuck" to the string and pulled it out from the center. Equal tugs to all sides pulled out the loop into a perfect circle.

CHECK IT OUT! Repeat this experiment but this time replace the thread loop with an arrangement of toothpick halves.

3.8 UMBRELLA IN A JAR

*I*t's raining, it's pouring.... You know how it goes. This experiment, however, isn't about bumping heads and going to bed. It's about surface tension and how it relates to umbrellas.

Materials
* *small jar*
* *cotton handkerchief*
* *rubber band*
* *water*
* *sink*

To Do
Fill a jar halfway with water. Pull a handkerchief across the mouth of the jar. Make sure that the fabric is pulled taunt. Then secure it with a rubber band.

Gently and slowly, turn the cup upside down over a sink. What happens to the water?

The Science
Cotton, like other fabrics, has tiny holes. When the jar was turned upside down, the weight of the water should have caused it to flow through the fabric's holes. It didn't.

Although you couldn't see it, the holes were patched by water molecules. Surface tension formed waterproof plugs across the fabric's holes. This skin was strong enough to prevent water from flowing through the porous fabric.

CHECK IT OUT! Umbrellas also depend upon water tension. Without its molecules hanging together, water would drip right through the holes of the umbrella's fabric.

3.9 PLUGGIN' THE HOLES

My roof leaks."

"Why don't you fix it?"

"I try, but there's a problem. When it's raining, I can see the leak but can't climb out onto the slippery roof to fix it. When it stops raining, I can climb out onto the roof but can't uncover the leak."

Materials

* plastic screen food basket
* pair of scissors
* bowl
* water

To Do

Fill a bowl halfway with water. Gently lower the food basket onto the surface of the water. Observe what happens to the level of water as the basket is lowered.

Now release the basket. What happens?

Use your scissors to enlarge one of the holes by making four small snips into the connecting plastic cross pieces. Does the basket still float? If so, enlarge the hole by snipping away one more of the plastic intersections.

Lower the basket onto the water. Keep snipping off pieces until the basket can no longer float.

The Science

Water molecules formed a tough skin across the holes of this basket. The skin was so tough that it kept water from flowing into the basket. As the hole size was increased, however, the surface tension was stressed. Eventually, the holes became too large to keep out the water. When that point was reached, water flowed through the hole and entered the basket. Within the basket, the water's additional weight offset the balance and the basket sank.

CHECK IT OUT! What happens to your floating basket when a drop of detergent is placed inside the basket? Any guesses?

3.10 WATERY HAND LENS

Suppose you examined a droplet of water from the side. What shape would you see? The bottom of the droplet would be as flat as the surface it sat upon. The upper surface, however, would take on the shape of a curve. It's this same type of curve that forms an optical lens. In fact, it works just the same. Want to see?

Materials
* paper clip
* water
* newspaper

To Do

Unbend a paper clip so that each of the two "U"s extends out in opposite directions into an "S" shape. Now bend up the free end of the smaller "U" so that it forms a tiny loop as shown in the illustration.

Dip this loop in water. As you slowly lift up the clip, a bead of water should remain within the loop. Describe the shape of the bead's upper and lower surface.

Place the bead several inches above a newspaper. Try reading the letters through the bead of water. You may have to adjust the height to get the correct focus. What happens to the letters?

Suppose you used a larger loop. How would that affect your observations? Make a guess and then try it to find out.

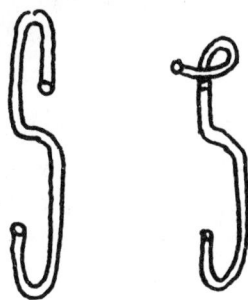

3.11 SOAP POWERED FISH

er 100 years ago, "motorized" fish were sold as cheap toys. These fish were constructed from paper. According to the instructions, all you had to do was add a drop of the right solution and off went the fish. But what was the right solution? Cod fish oil?

Materials
* small bowl
* pair of scissors
* water
* heavy stock paper
* dishwashing liquid

To Do
Fill a small bowl with water. Use a pair of scissors to cut out the fish pattern shown below.

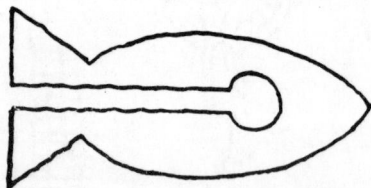

Place the pattern in the center of the bowl. Observe its movement. Now place a drop of dishwashing liquid within the center circle. What happens now?

The Science
When the paper fish was placed in the bowl, it floated on water particles that pulled equally in all directions. These tiny forces canceled each other out and the fish remained still.

When a drop of dishwashing liquid was added, the balance of surface tension was destroyed. As the soap flowed out the back of the fish, it broke up the surface tension. Since the pulling force at the front was still intact, the fish moved forward.

CHECK IT OUT! Try angling the cut to one side of the fish. How does this affect its movement?

3.12 BUBBLES

*H*ave you ever played with a magnet and a paper clip? If so, you know that as the magnet and clip are separated, the magnetic attraction decreases.

Now imagine the tiny forces between neighboring water particles. While they are side by side, their attraction is strong. But if we separate the particles, the attraction is lessened. Enough science—let's have some fun.

Materials
* cup
* dishwashing liquid
* measuring cup
* pipe cleaner
* water

To Do
Fill a cup halfway with a mixture that is made from one part dishwashing liquid and nine parts water. Gently, but thoroughly, mix this solution.

To form a bubble wand, bend one end of a pipe cleaner into a large loop. Dip the bubble wand into your solution. Remove the wand and a film of bubble solution will stretch across the loop opening.

As you gently blow into this loop, bubbles will emerge from the opposite side.

The Science
When dishwashing liquid is added to water, the liquid breaks down into very tiny particles. These particles of dishwashing liquid spread out and get between neighboring water molecules. There, they interfere with the attraction between water molecules. This allows the water to spread apart into a thin film. Without the interference of detergent, the intense attraction between the water particles would "snap" the shape shut and not allow the skin to be drawn into a film. When this film surrounds a pocket of air, it becomes a bubble.

3.13 BUBBLE DOMES

ere's a bubbly challenge. Can you place a bubble within a bubble? Perhaps. But can you place a third bubble within second...and a fourth within the third?

How many bubbles within a bubble can you create? There is a rumor that some kid from Kalamazoo can create a giant bubble that contains four levels of smaller bubbles. Do you think you can break the Kalamazoo Kid's record?

Materials
* dishwashing liquid
* water
* sink
* cup
* straw

To Do
Make a bubble solution using one part dishwashing liquid and nine parts water. Dip the end of a straw into the bubble solution.

Wet the surface of a sink with a layer of bubble solution. Place the wetted end against the wet sink surface. Gently blow into the dry end of the straw. Once the dome bubble has been made, withdraw the straw. The bubble seals itself and remains stable.

Now rewet the straw and inject it into the bubble. Once the tip touches the sink surface, blow your next bubble. Keep going until you approach or break the Kalamazoo Kid's record.

3.14 BUBBLE FRAME

*D*id you know that your hand can be a great bubble maker? Just make an "okay" sign with your thumb and index finger. Soak this "loop" into a bubble solution. Then blow into it gently and steadily.

Neat, huh? Well, here's another type of bubble maker. This one is called a bubble frame.

Materials
- *kite string*
- *pair of scissors*
- *two straws*
- *cup*
- *dishwashing liquid*
- *shallow pie pan*

To Do
Make a bubble solution using one part dishwashing liquid and nine parts water. Pour this solution into a shallow pie pan.

Cut a piece of string that is about four times the length of a straw. Thread the string through two straws. Knot the free ends of the string forming a closed loop.

Soak the entire loop in the bubble solution. Remove the loop. Slowly spread apart the straws so that a rectangular frame forms.

Gently blow onto the film that stretches across the frame opening. When the bubble has formed, pinch it off by closing the frame.

The Science
The soaked string and straw serve as a reservoir for the bubble solution. As you blow onto the film, the stretched "skin" of bubble solution expands. When the frame is closed, the bubble film sticks back onto itself and pinches off from the plastic straw and cotton string.

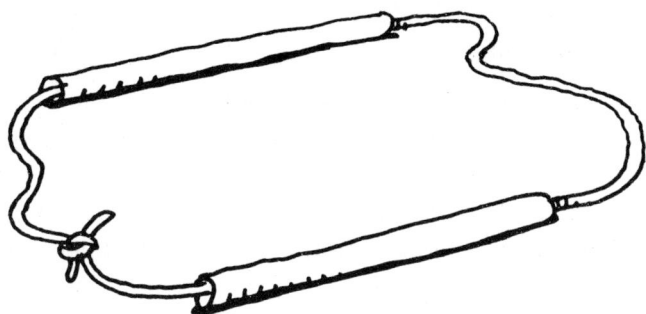

3.15 ROPE BUBBLE MAKER

If you want to make the biggest possible bubble, there are several things to keep in mind. First, the biggest bubbles form in humid surroundings. So try to avoid the dry air of winter.

If, however, you're stuck with cold weather, you may want to move your bubble making operation to the humid surroundings of a bathroom (with the shower turned on).

Also, direct sunlight is bad for bubbles. Try making them in the shade. And, of course, you'll need a special bubble maker such as the one described below.

Materials
* 3-foot-long cotton rope
* pair of scissors
* large plastic cup or container
* dishwashing liquid

To Do
Make a bubble solution from one part dishwashing liquid and nine parts water. Pour this solution into a large plastic cup or container.

Cut a length of cotton rope about 3 feet long. Knot the free ends of the rope to form a large loop. Immerse the loop into the bubble solution. Keep it in the solution for several minutes so that the entire rope becomes saturated with the liquid.

Remove the rope from the solution. Because it is saturated, it will sag. Position your hands on opposite sides of this closed loop. Gently and steadily spread the loop apart. The bubble film should span the opening.

Slowly walk with the loop extended. The passing air will enter the loop and inflate the bubble. When the bubble has grown to a large enough size, close the loop so that the bubble gets pinched off.

CHECK IT OUT! Try making a giant bubble making loop from a length of rope that's at least 6 feet long. For this bubble maker, you'll need to use a large container filled with bubble solution and a friend to help "work" the loop.

THE PUSHES AND PULLS OF AIR

4.1 AN INVISIBLE OCEAN

*L*ook around you and what do you see in the space between your eyes and this page? Hopefully, nothing. But seeing nothing, doesn't mean that nothing is there. Actually, something very real fills this space. That something is called air. Even though it is invisible, air is there.

Materials
* clear plastic cup
* large transparent container, such as a fish bowl or aquarium tank
* water
* sheet of paper

To Do
Fill the large container halfway with water. Crumple a sheet of paper into a small ball. Place this crumbled ball into the bottom of an empty plastic cup. Turn the cup so that it is upside down. Push the cup into the water. Look at the inside of the cup. What do you see? What happens?

Remove the cup from the container. Take out the crumpled paper. Is the paper wet? Can you explain your observation?

Stuff the dry paper back into the cup bottom. This time hold the cup so that the open end faces upwards. Slowly, lower the cup into the water. What happens this time? Can you explain why?

The Science
Although you can't see it, we are surrounded by an ocean of air. When the cup was first pushed upside down into the tank, air became trapped in the cup's space. Since the cup was filled with air, there was no room to allow water into the cup.

When the cup's opening faced upwards, the trap was open. As it was lowered into the liquid, water flowed over the brim and pushed the air outward into the surrounding atmosphere. Since the flow of water wasn't stopped, the paper got wet.

4.2 POURING AIR

Imagine pouring a glass of water. As you tilt the container, the water rises to the brim. Soon it overflows the lip of the glass. As it flows over the edge, the water forms a steady downward stream. DOWNWARD, as in going down towards the floor. But did you know that some fluids pour upwards?

Materials
* two clear drinking glasses
* aquarium or large fish bowl
* water

To Do
Fill an aquarium tank 2/3 full with water. Place one glass in the tank so that it fills completely with water. Then turn it upside down.

Take a second glass, hold it upside down, and push it into the water. What happens to the glass? Does it fill with water?

Position this air-filled glass near the bottom of the tank and directly beneath the water-filled glass. Tilt the air-filled glass so that a bubble of air escapes. Where does the air go?

Continue tilting the glass so that its air contents flow upward and into the other glass. What happens to the water-filled glass as air enters it? Can you explain this observation?

The Science
When the air-filled glass was tilted, the air escaped as rising bubbles. Instead of reaching the surface, the stream of rising bubbles were "intercepted" by the opening of the water-filled glass. There, the rising air became trapped and filled this second glass.

4.3 DOWN THE DRAIN

A simple can opener is a great device for explaining levers. However, we won't talk about its use as a simple machine in this experiment. For now, think about using a can opener to punch one hole near the rim of a small liquid-filled can. Turn the can on its side and see what happens. Not much. Although there's a hole in the container, the liquid doesn't seem to flow out. Something is in the way!

Materials
- *funnel*
- *clay*
- *cup*
- *pencil*
- *clear plastic beverage container*

To Do
Place a roll of clay around the mouth of a plastic container. Insert a funnel in the mouth and press the clay firmly against the spout to form an airtight seal. Slowly pour water from a cup into the funnel. What happens? Can you describe the movement of two fluids, even the one you can't see?

Empty the container and place another roll of clay along its mouth. Place a small plug of clay in the spout of the funnel. Insert the funnel back into the container and use the clay roll to form another airtight seal. Fill the funnel to the brim (the clay plug should prevent water from leaking out of the funnel) with water. Use a pencil to poke out the clay plug. What happens? What forces are at work?

The Science
In the first situation, water that was poured into the funnel swirled around its inner surface and fell into the container. As it moved into the container, it did not fully block the spout opening. This opening offered an escape route for the trapped air. Water went into the container. Air went out of the container.

Things changed when a clay plug was used to block the spout. When the plug was knocked out, no solid material blocked the flow of water.

There was, however, a container full of trapped air which kept the water out. Since the water filled the entire spout opening, there was no escape path for the air.

4.4 UNDERWATER BOAT

Blub. Blub. Blub. Think of an underwater boat and most likely two images come to mind: submarines and sunken wrecks. In this experiment, you'll construct a third type of submerged vessel. Unlike the other two, this one will actually "float" beneath the water's surface. Interested?

Materials
- *clear plastic cup*
- *large transparent container, such as a fish bowl or aquarium tank*
- *water*
- *aluminum foil*
- *toothpick*
- *paper*
- *clay*

To Do
Build a small boat out of aluminum foil, paper, and other construction materials. Float this boat on the surface of the water-filled fish bowl.

Place an empty and clear cup over the boat and gently push down. What happens to the boat as the cup is pushed into the water? Does the boat continue to float? Why? What happens to the level of water in the cup?

The Science
When your boat was placed in water, it floated on the liquid's surface. The cup that was placed over the water was filled with air. As this cup was pushed into the water, its trapped air pushed down on the surface below. This push was enough to drop the water's surface. So as the cup went down, so did the boat which still remained at the surface of this trapped water pocket (although this surface was now underwater).

✳ 4.5 FLYING STRIP

*A*s the autumn wind rips down the street, leaves seem to fly into the air. Although these discarded plant parts lack wings and engines, they seem to have little difficulty in becoming airborne. Why? What makes leaves so special? Is it their shape, or is something else at work?

Materials
* *2 inch × 11 inch strip of paper*

To Do
Make a fold near the end of a strip of paper. While holding this folded edge, the rest of the strip should fall freely. Now blow at the bottom of the strip. What happens to it?

Hold the strip right beneath your bottom lip. Make a prediction. How will the strip behave if you blow now? Blow outward with a long steady stream of air. What happens to the strip?

The Science
When you blew at the bottom of the strip, the paper went up. That seemed logical. The force of your airstream "pushed" against the bottom of the paper, causing it to rise from its hanging position.

However, when you blew across the top surface of the paper, it also went up. This odd "rising" property is described by a scientific effect called *Bernoulli's Principle* (we'll call it BP for short). BP states that as air (or any fluid) moves, it creates a region of low pressure. Low pressure air has less "pushing force" than air of higher pressure.

Therefore, when you blew across the upper surface of the strip, you created a region of low pressure. At the same time, air that was beneath the strip kept up its normal pressure. Since the normal pressure was now greater than the reduced pressure up top, it pushed the strip upward.

4.6 LEVEL CLIMB

*O*kay, so you're thirsty. Stick a straw into the glass full of soft drink and suck up. As you've observed a thousand times before, the soft drink rises up the straw and enters your mouth. It's a pretty neat transport system. But did you know that you can get the soft drink to climb the straw by blowing out?.

Materials
- *long straw*
- *water*
- *drinking glass*
- *pair of scissors*

To Do
Snip the straw with a pair of scissors but make sure that you do not cut fully through the straw. Your cut needs to leave a small piece of straw intact so that after the cut, the two halves stick together. The piece that attaches the two straws should be flexible.

Fill a glass ⅔ full of water. Place the modified straw in the glass. Bend the two sections of the straw so that they form a right angle.

Blow through the horizontal straw mouthpiece. Observe the level of liquid in the vertical section. What happens to the liquid as you blow into the straw? What causes this change? How is this like sucking up liquid through a straw? How is it different?

The Science
As you blew through the horizontal straw section, you created a fast moving blast of air. This moving air created a region of lower pressure that was centered above the vertical straw's opening. The surrounding air still pressed down with the same pressure on the exposed liquid surface in the glass. The pressure at the top of the straw, however, was lessened.

This imbalance of force pushed down on the water's surface, causing the liquid to climb up the straw. When you suck on a straw, you create the same imbalance, allowing the atmosphere's pressure to push the liquid up the straw!

4.7 BLOWN AWAY? ...PERHAPS

Examine a weather map and you're likely to uncover a capital "H" and "L" placed somewhere over the terrain. The "H" identifies an air mass with high pressure. The "L" identifies a low-pressure air mass. Big deal. Actually, it is a big deal. Not only can you predict weather changes from this information, but you can learn more about winds and their mostly likely direction and strength.

Materials
* two table tennis balls
* tape
* two 1-foot-long threads
* ruler

To Do
Break off two lengths of thread about 1 foot long. Use these threads and tape to attach two table tennis balls to a ruler. The balls should be placed so that when hanging freely, they are 2–3 inches apart (not difficult to accomplish when you're taping them to a ruler scale!).

Hold the table tennis balls about a foot in front of your mouth. Make a guess. If you blow an airstream directly at the gap between the balls, how far will the balls separate. After you've made your guess, check it out.

The Science
Wow. They didn't separate! Instead, the balls acted as if they were attracted to each other. What gives? It's that low pressure phenomenon again. The airstream that moved between the balls had lower pressure than the surrounding air. This difference in pressure caused the surrounding air to push inwards on the stream. Since the balls were in the way, they moved into the stream of flowing air.

4.8 SPIN BY ME

Ever hear of a safety plug? It's a device that is found at the end of the power cord of some home appliances, such as hair dryers. It looks like a box and has prongs that go into a wall outlet. Look closely and you'll see that the safety plug has a small bottom, or switch, labeled "reset."

If a hair dryer accidentally gets wet, its electrical circuits can short out. A short circuit produces a huge surge of current. The safety plug detects this extraordinary current flow and shuts off the electricity before someone gets a serious shock. Later on, you can switch the safety plug back on by pressing the reset switch. As you might imagine, these plugs are very important safety devices. Never use a hair dryer that is not equipped with a safety plug...especially in this next experiment!

CAUTION
Do not use a hair dryer or any other electrical device near water. Remember, if used inappropriately the hair dryer can easily get hot enough to burn you!

Materials
* table tennis ball
* hair dryer (with safety plug)

To Do
Take your hair dryer and table tennis ball into the living room. Perform this experiment away from any source of water (including a fish tank).

Switch the settings on the hair dryer to "cool." Do not use any of the dryer's hot or warm air options. Keep it cool!

Carefully plug the safety plug end of the power cord into an electrical outlet. Turn on the hair dryer and hold it so the blast of air shoots straight upwards.

Gently place a table tennis ball in this air stream. What happens? How far is the ball shot? At what angle will the stream "lose" the ball?

The Science

The ball floats on the jet of fast moving air whose motion produces a region of low pressure. Since the surrounding air has a slightly greater pressure, it pushes inward on the stream from all directions. This inward force is strong enough to "trap" the table tennis ball within the air jet. If, however, the stream is tilted too much, the pressure difference cannot overcome the pull of gravity and the ball falls from the stream.

4.9 PICK-UP GAME

When you apply a force to something, that object moves in the same direction as the force. Push a toy car, and the car rolls in the direction of the push. Hit a ball with a bat, and the ball flies off in the direction of the hit. It's basic logic. It's also one of *The Laws of Motion*.

In this next experiment, you may observe something that goes against this law...or does it? Can you figure out what forces account for the odd behavior of this table tennis ball?

Materials
* table tennis ball
* small funnel

To Do
Make sure that the funnel has been cleaned with soap and rinsed thoroughly. Hold the funnel face down.

Place a table tennis ball into the funnel end. Gently hold the ball upwards against the spout opening. While holding it in place, blow out a long steady stream of air.

Now release the ball. What happens? What forces are at work? How long can the ball stay up?

The Science
As you blew into the funnel, it created a rush of fast-moving air. This air produced a region of low pressure. Since the ball was "bathed" in this low pressure current, forces other than its weight needed to be considered.

If the pressure difference is great enough between the air stream and the surrounding air, the force overcomes gravity and can lift the ball. Although there is still a downward force "blown" onto the ball, the inward push due to the difference in air pressure is the winner in this competition of forces.

4.10 FLY BALL GAME

Skeeball is a game in which you must land a wooden ball into a hoop. You start by rolling the ball down a ramp. At the end of the ramp, the ball becomes airborne. If your aim was accurate, the ball lands within a small hoop. If you miss your target completely, the ball rolls back down the slope with little or no scored points.

In this experiment, you'll get to assemble a similar game. You won't, however, roll the ball with your arm. Instead, take a deep breath as you read the rules.

Materials
- *table tennis ball*
- *several small cups*
- *marker*
- *tape*

To Do
On a desktop, place several cups at various distances away from the "launch" cup. Secure the cups to the desktop with tape.

Give each cup a point value based upon its distance from the launch cup. Write this value on the cup. The closest cup should have the lowest point value while the farthest cup should have the greatest point value.

Place the table tennis ball in the launch cup. Direct a blast of air into the cup at an angle so that it will carry the ball towards a target cup. Alternate turns with an opponent. If you miss all cups, you don't receive any points. If you land the ball in a cup, you score the value of points listed on the cup's side.

The Science
It's that low pressure phenomenon again. The stream of moving air creates a region of low pressure. As the ball pops out of the launch cup, it gets "trapped" within the low pressure flow and was kept in place by the surrounding high pressure air. As gravity takes its toll, the ball drops—ideally into the cup with the highest point value!

4.11 MAY THE FORCE BE WITH YOU

Okay, it's time to dig into your old toy chest. Forget about the broken cars, doll arms, and game pieces. We're looking for those plastic darts with suction cups. You know the type. They are the ones that never seen to stick on the target—although they seem to do well on your younger brother's forehead.

Materials
- *two suction cup darts*
- *thumbtack*
- *tape*
- *playing card*
- *water*

To Do
Make a hole through the center of a playing card with a thumbtack. Use a small piece of tape to cover one side of this hole.

Push a suction cup dart against the playing card so that it covers the untaped side of the hole. What happens?

Moisten the "lip" of the suction cup. Try sticking the cup against the card again. Does wetting the suction cup affect the way it sticks?

While the cup is stuck to the card, remove the tape. What happens to the suction cup if the tape is peeled away. Can you explain your observations?

The Science
The air that surrounds us pushes out with a force we call air pressure. The amount of air pressure is related to the concentration of air particles. The greater the concentration of particles, the greater the pressure. The fewer the particles per space, the lower the pressure.

When the suction cup is pushed against the card, air is forced out of the cup. Since air leaves this space, there are fewer particles trapped within the cup. As the cup "unsquishes," it produces a slight drop in the concentration of air particles. Hence, there is a lower air pressure inside the cup.

Outside the cup, however, things have stayed mostly the same. Since the surrounding air has a greater concentration of air particles, it has a higher pressure. This pressure pushes in the cup and keeps it from moving off its attached surface.

When the tape was pulled off the card, air rushed into the low pressure chamber. As the concentration of air particles equalized, the force keeping the cup stuck to the card dwindled.

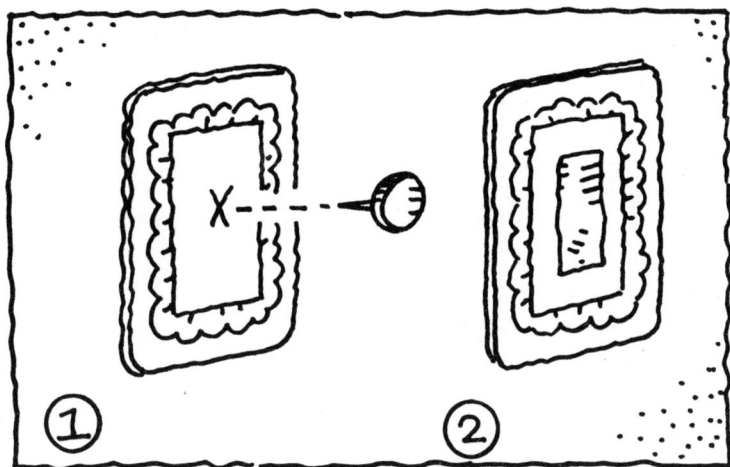

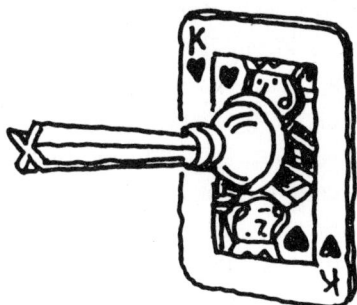

4.12 UPSIDE-DOWN MAGIC

e live our lives surrounded by an ocean of air. Although we aren't aware of it, this air has weight. The weight pushes down, in, out, and up on everything. Even though we can't feel this *air pressure*, we can certainly observe its effects.

Materials
- *paper*
- *piece of cardboard*
- *paper cup*
- *thumbtack*
- *water*
- *sink*

To Do
Fill a paper cup to the brim with water. Then carefully place a small square-shaped piece of cardboard over the mouth of the cup.

Place one finger in the center of the cardboard. Then, while holding the cup over a sink, slowly turn the cup over. Remove your finger and the cardboard remains in place.

Gently punch a thumbtack hole into the bottom of the cup that is now facing upwards. What happens?

The Science
Under the weight of the water, the cardboard bottom sags slightly. This "drop" produces a slight drop in concentration of air, which is trapped in the cup. This drop produces a region of low pressure. Since the outside air remains at the same concentration, it retains a relatively greater pressure than the air inside the cup. This pressure difference is enough to keep the card in place even though it is supporting a cup full of water!

When a hole was punched into the cup, air rushed in to equalize the pressure. As the pressure balanced, the card fell away under the water's weight.

4.13 THE STRAW DROPPER

*H*ave you ever misplaced a medicine dropper that was needed for an awesome experiment? If so, don't worry. Here's how you can use a straw to make a *precision liquid delivery system*.

Materials
* *straw*
* *cup filled with water*
* *sink*

To Do
Insert a straw into a cup filled with water. Place your finger over the straw's open end, making an airtight seal. Then lift the straw out of the water. Place the straw over a sink and release the seal. What happens?

Now place your finger over one end of an empty straw to form an airtight seal. While keeping this seal, insert the straw into a cup filled with water. Wait several seconds. Remove the straw and place it over a sink. Release the seal. What happens now?

The Science
When you immersed the open-ended straw into the water, it filled with liquid. By sealing the top of the straw, you formed an airtight chamber within the straw. When the straw was lifted out of the water, the liquid within the straw sagged downward (you might have observed the curved drop at the straw's bottom). The slight drop was enough to create a pressure difference between the chamber and the outside surrounding air. That air kept pressing against the liquid in the straw to prevent it from spilling out.

When you immersed the closed-ended straw into the water, it did not fill. By sealing the end, you created an enclosed air space. This chamber of air exerted a pressure that kept water from entering up the straw. Since no water could get in, the straw remained dry.

4.14 WATER TOWER

Although you can't see it, the surface of a glass of water is under plenty of pressure. The pressure is formed from the weight of a column of air that stretches upwards for over 20 miles! Although the water's flat and calm surface may not offer clues to this unseen force, here's a way for you to observe its effects.

Materials
* clear plastic container
* plastic soft drink bottle
* water

To Do
Fill the bottle with water. Position it above the empty container. Invert the bottle so that the water flows into the container. Keep the mouth of the bottle positioned about 1 inch from the bottom of the container. What happens when the water level rises to the level of the bottle's mouth?

Raise the bottle to 2 inches from the bottom. What happens now?

Refill the bottle until the container is 1/3 full of water. Cover the bottle's mouth to form a water-tight seal. Invert the bottle and place its mouth below the water line. Remove your hand. What happens to the level of water within the container? Can you explain your observation?

The Science
When the bottle is turned upside down, the weight of the water causes this fluid to flow downward. It keeps flowing until the water level rises to the level of the bottle's mouth. At this point, the level of the rising water "shuts off" the bottle's flow. Here's why:

The weight of the surrounding air pushes down on the surface of the rising water. This downward push is transferred in all directions through the particles of water. The force that is transferred in the upwards direction resists the water spilling out of the bottle.

Since the atmospheric force counteracts the weight of the water, it stops the downward flow and prevents additional water from spilling out of the bottle.

Although the tower of water supported in this activity was only a few inches tall, standard air pressure can keep up a column of water that is higher than a 3-story building!!! (34 feet to be exact.)

4.15 SIPHON

ere's an adventure that doesn't require any work—just as long as you set it up correctly. Then sit back and watch as the physics of flowing takes control.

Materials
* two containers
* a length of plastic hose or aquarium air pump tubing
* chair

To Do
Fill one container ¾ full with water. Place this container on the flat seat of a chair. Place the empty container on the floor in front of the chair.

Fill the tubing with water. Stop both ends with your fingers. Place one end of the tubing in the lower container. Place the other end in the upper container.

Remove your fingers from both ends. What happens?

The Science
Congratulations! You've created a siphon. It is a bent tube with arms of unequal length. Liquid travels through the siphon from one container to the other until the levels of liquid in both containers are the same.

Air pressure pushes down on the surface of the liquid that is found in both containers. The force of this downward push supports a column of fluid in the arm directly above this liquid. However, the longer arm contains a greater "weight" of fluid pushing down than the smaller arm. The arm with the greater weight overcomes the upward resistance (due to air pressure) and the liquid slides down this arm. As it moves into the lower container, the liquid draws up fluid through the siphon's shorter arm.

4.16 PRESSURE DROP

In this experiment, you'll be punching a pencil through a milk carton. Just remember to empty the carton before you start doing this. Otherwise, this adventure will get pretty messy.

Materials
* milk carton
* pencil
* sink
* tape
* water

To Do
Rinse out an empty milk carton. Use a sharp pencil to carefully poke two holes in the carton. The holes should be positioned one on top of the other and several inches apart.

Use tape to cover up the holes. Fill the carton with water and place it next to a sink so that the openings aim into the sink basin. Pull off the tape. How do the two jets compare? Can you explain your observations?

The Science
To understand the difference in jets, let's take a look at the amount of water that is found above each hole. The hole that is near the container's top has less water above it than the corresponding hole below it.

Less water produces less weight. Less weight results in a lower pressure. Therefore, the jet near the top of the container is under less pressure. Hence, its stream is not as energetic as the jet that emerges from the lower (higher pressure) hole.

4.17 PAPER PRESSURE

Is a sheet of paper strong enough to break a wooden ruler? You might be surprised to discover the answer—and since this answer might ruin your ruler, you might want to try this with a wooden paint stirrer.

CAUTION
To protect your eyes, you should only perform this experiment when wearing safety goggles.

Materials
* two wooden paint stirrers (available at paint or hardware stores)
* sheet of newspaper
* hammer
* safety goggles (available at hardware stores)
* table

To Do
Place a wooden paint stirrer on a sturdy table. Position the stirrer so that it extends several inches past the table's edge. Place an opened sheet of newspaper over the half of the stirrer that is on the table-top.

Put on a pair of safety goggles. Use the hammer to carefully strike the end of the stirrer that extends over the table's edge. What happens to this piece of wood?

Repeat this experiment. This time, however, fold the newspaper into a thin rectangular bundle. Place this bundle over the half of the paint stirrer that is on the table. What happens when the stirrer is struck this time?

The Science
Air pressure pushes down with a force of about 15 pounds per square inch. To determine the total force, you need to multiply the exposed surface area by the 15-pound/inch force. As you can see, a spread out page has a much great surface area than the folded page. Therefore, a much greater pressure pushes down on the unfolded page.

When the hammer slammed against the hanging end of the stirrer, the opposite end moved against the downward force of air. When this end was covered by an unfolded page, the air pressure produced enough resistance to break the stirrer. When a folded page was used, less air pressure resisted the stirrer's movement.

4.18 GOING IN?

ill a balloon with air and knot it. The balloon stays about the same size because there is a balance of forces. The force exerted by the air inside the balloon equals the force pushing in by the atmosphere. Since things are equal, the balloon stays the same size.

In this experiment, you'll see what happens when forces are unequal.

Materials
* 1-liter soft drink container
* balloon
* warm water

To Do
Run a stream of warm water over the outside of a 1-liter plastic container. After about a minute remove the container from the stream. Stretch a rubber balloon over the opened mouth of the container and allow it cool. You can hasten the cooling by placing the container in cool water. What happens to the balloon?

The Science
As you warmed the container, you energized the air particles that were inside the container. When heated, the particles moved faster. As they moved, many of them escaped through the neck of the container. Then the passageway was sealed when you stretch the balloon over it.

As the container cooled, the trapped air particles lost energy. Since the air particles were not as energetic, they moved slower. There were also fewer of them (don't forget that many escaped).

This resulted in a greater concentration of air on the outside than the inside of the container. The difference in concentration produced a similar difference in pressure. This pressure difference pushed the balloon into the container.

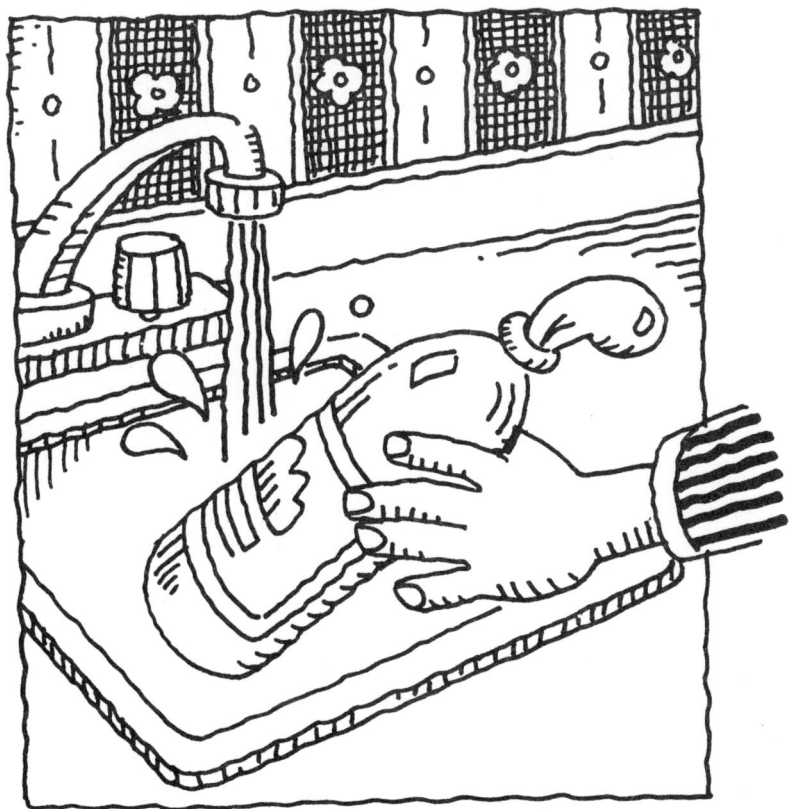

4.19 CAN CRUSHER

*I*magine a small, empty can. On its outside, thousands of pounds of pressure are constantly pushing inward. But what is inside the can? Air.

The air that's in the can is also at atmospheric pressure. It pushes back with equal strength against the crushing force. The result is that nothing happens. The forces cancel each other out and the can remains as if no forces were active at all.

But what would happen if we were able to get rid of the inside air?

Materials
* large empty olive oil can (with cap)
* high wattage hair dryer with safety plug
* two kitchen oven mitts

CAUTION
This experiment uses high temperatures and should only be performed with adult supervision. Use only a thoroughly cleaned olive oil can and perform the heating while wearing oven mitts.

To Do
Thoroughly wash an empty olive oil can with a soap water solution. Pour out the solution and rinse well. Make sure to wash any oil from the inside of the cap as well. Turn the can upside down and allow it to dry overnight.

Place the clean, dry can on a protective surface far from a sink or running water. Put on a pair of oven mitts. Use a high wattage hair dryer to warm the outside of the can. Keep the nozzle of the dryer several inches from the can's surface. Uniformly heat the can for several minutes and then shut off the dryer.

With the oven mitts still on, very carefully screw on the cap of the olive oil container so that it becomes airtight. Now sit back and watch.

The Science
As you heated the can, you energized the particles of air that occupied the inside space. As these particles gained energy, they moved more

quickly. Many of these "energized" particles gained enough energy to escape through the can's opening. This resulted in fewer (although highly energetic) particles of air within the can.

As the can cooled, the trapped air particles lost their energy. They could no longer produce enough collisions to balance the air pressure pushing in from the outside. Since the inner pressure could not counteract the outside air pressure, the can collapsed.

4.20 UNSEEN CRASHES

Decades ago, space explorers performed an experiment on the surface of the moon. In this airless environment, an astronaut released a hammer and a feather at the same time. As predicted by the law of gravitation, both objects fell side by side and struck the lunar surface at the same time.

You won't need to travel to the moon to test this experiment on gravity.

Materials

* *two sheets of paper*

To Do

Hold a sheet of paper horizontal. Release it and observe its fall to the floor. Describe its descent. Did it fall in a straight line or did it seem to sway back and forth?

Crumple a second piece of paper. Release both the flat sheet and crumpled paper. Compare their falls. How did crumpling the paper affect its motion? What might account for the change?

The Science

All objects should fall to the Earth at the same speed. They don't. The difference is due to something called *air resistance*. Air resistance is produced by the collisions an object makes with invisible particles of air.

As the object collides with air, the air slows it down. Shapes that are round or streamlined encounter less resistance than flat, broad shapes. The flat sheet of paper collided with many more air particles than the crumpled sheet. This caused the flat sheet to fall much more slowly than the sheet that was crumpled into a ball.

CHECK IT OUT! If both the crumpled and flat sheet of paper were dropped at the same time on the moon, they'd hit the lunar surface at the same time.

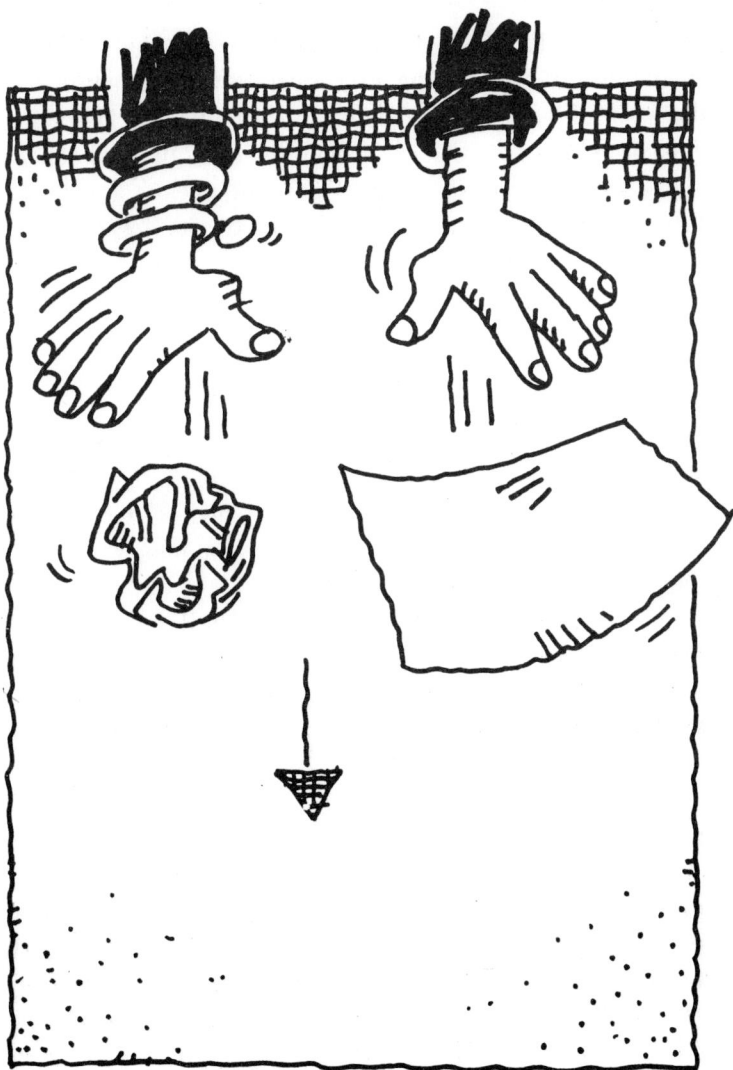

4.21 OUT OF THE WAY

Hey, good buddy, it's a convoy! Somewhere on this side of interstate 70, five trucks are forming a tight procession. Although it might look like a social way to drive, convoys save fuel. By keeping close behind each other, the trucks encounter less air resistance and require less gasoline. The lead truck, however, doesn't get the advantage since this vehicle "parts the air" for the trucks that follow.

Materials
* *a sheet of paper*
* *a book*
* *pillow*

To Do
Find a book that is slightly larger than your sheet of paper. Hold out the book so that it lies horizontal to the floor. You may want to place a pillow beneath the book to cushion its fall.

Position a sheet of a paper on top of the book. Before releasing the book, predict what will happen to the paper as the book drops away. Will the paper stay with the book? Will the paper slowly separate from the book and sway to the floor? Make a guess and then release the book.

The Science
Instead of swaying back and forth, the paper dropped as quickly as the book. This increase in speed was due to decreased air resistance.

As the book fell, it moved air particles aside. Since these particles were moved out of the paper's way, they could not slow the sheet of paper. In addition, since air could not get between the paper and the book, a suction was created that keep the paper and book together.

CHECK IT OUT! Bicycle racing teams also travel in convoylike formations. Can you explain why?

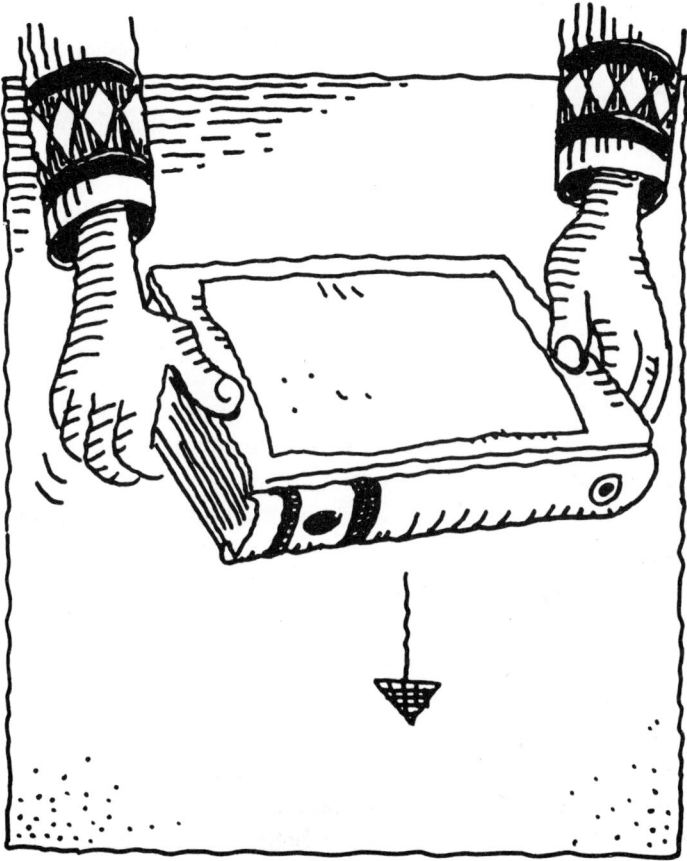

4.22 HOOKING UP

*I*magine Galileo standing at the top of the Leaning Tower of Pisa in Italy. He is holding two different-sized spheres. He releases them at the same time. As expected, they strike the ground at the same time. One is large. The other is small. Yet they fall to Earth at the same rate. It makes sense, or does it?

Materials
* *three identical rubber balls*
* *tape*
* *6-inch-long string*
* *a pair of scissors*

To Do
Extend your arms while holding out two separate rubber balls. Release the balls at the same time. Both balls should fall to the floor at almost the same instant. This should make sense because they both contain the same amount of "atomic material" (also known as matter).

Now let's make a falling body of twice the mass. To build this object, cut a section of string about 6 inches long. Tape one end of the string to each ball. This string connector makes this object a "double" ball. Some scientists refer to this complex mass as a *system*.

Think about dropping it. Should the system fall at a faster rate than a single ball? Will one ball of the system fall faster than the other ball?

Suppose you held this system in one hand and a single rubber ball in the other hand. Imagine dropping both loads at the same time. Should the system fall faster?

If so, how does each ball of the system know to increase its speed. Is a message sent along the connecting string?

The Science
Whether it is a single ball or a system of two, the loads will fall to the floor at the same rate.

Think about it. There is no "telepathy" that tells the individual balls of the system that they should fall faster as a result of being attached by a string to another ball. Drop the system and neither component ball falls faster than the other.

Let's expand this idea. Imagine a five ball system. Although the five balls are attached by strings, this system will still fall at the rate of a single ball.

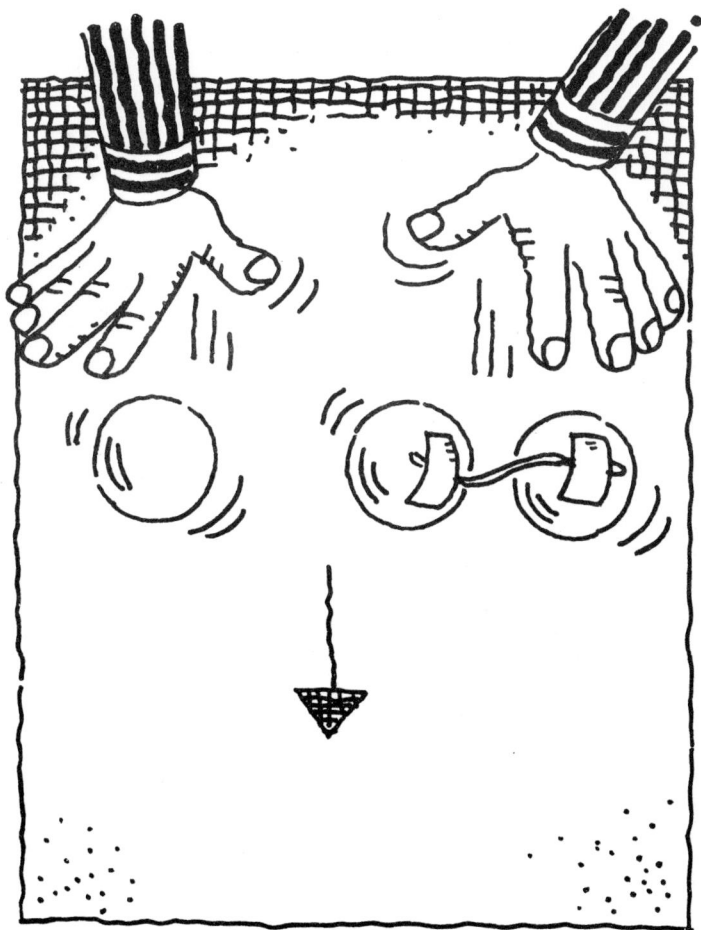

MOTION

MADNESS

5.1 KEEP CRANKING

When it comes to doing work, it often helps to have an advantage. That's where machines come in. Machines are devices that give us an advantage.

There is a trade-off, though. Machines are designed so that less force is needed to accomplish a task. The lessened force, however, must be applied for a longer time or over a greater distance.

Materials

* a couple of paper clips

To Do

Unbend a paper clip. Straighten it out as best as you can, but don't worry if it's not perfect.

Hold one end of the clip tightly between the thumb and index finger of your left hand. Grasp the free end of the clip with your right hand. Try to roll the clip using the fingers of your right hand.

Most likely, the clip didn't turn. Spinning a straight wire that is held in place can be hard to do. What you need is an advantage—a mechanical advantage!

Place two bends in the clip so it looks like this:

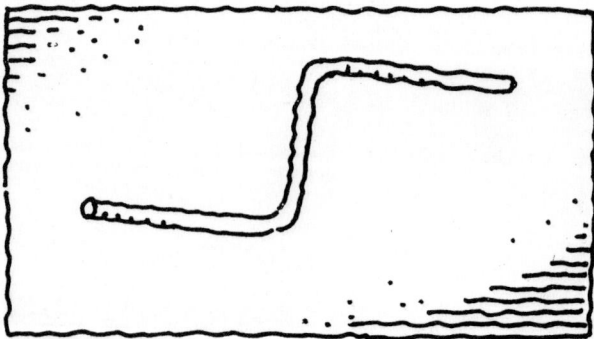

Now grasp one end of the clip. Hold the opposite end and rotate the clip as if it were a crank.

The Science

Congratulations. The clip has become a crank. You've built a simple machine! By changing the way the force was applied to the clip, you gained an advantage. This advantage made turning the clip easier.

CHECK IT OUT! Although it may not look like one, a doorknob is a type of circular crank. Its wide knob lets you apply a force that can turn the inner shaft. Without a knob, it's almost impossible to turn the shaft.

Next time you're in a hardware store, look for a doorknob display. If you're lucky, you'll get a chance to check out this advantage.

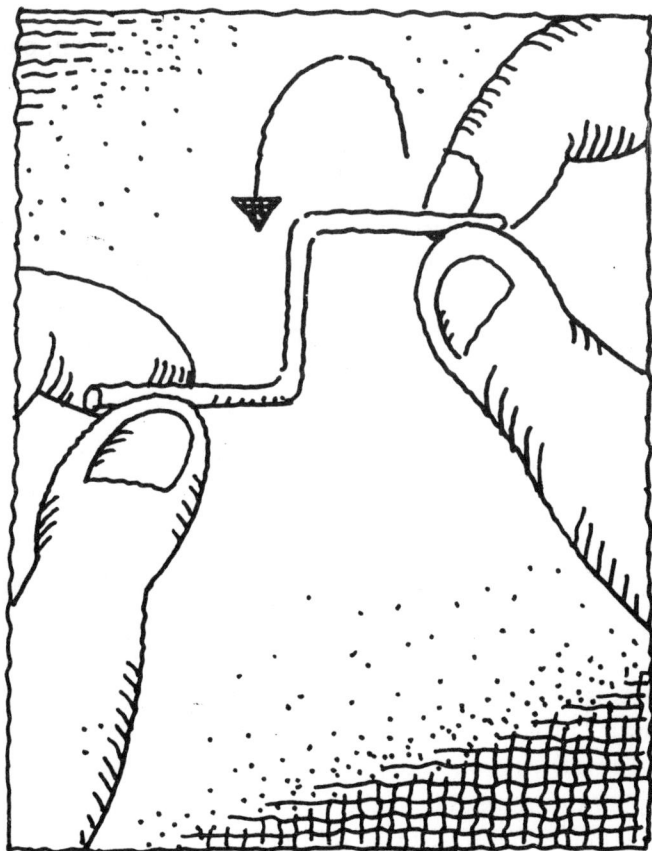

5.2 A SPIN ON THINGS

Have you ever watched a bulldozer move? If so, you know that it moves on two metal belts called tracks. When both tracks spin at the same rate, the bulldozer goes straight ahead. When the left rack spins faster, it creates an off-centered force that causes the machine to turn to the right. Likewise, when the right track spins faster, the machine turns to the left.

Materials

* empty milk carton
* 2-foot-long string
* water
* nail
* sink

To Do

Use the nail to punch five holes in an empty milk carton. One hole is centered on the top of the carton. The other four holes are placed to the left of each lower corner, as shown in the illustration.

Tie a 2-foot-long string to the top hole. Open the carton spout. Over a sink, quickly fill the carton with water. Hold it away from the faucet. What happens? Can you understand the connection between the water jets and the movement of the carton? How are these jets like the tracks of a bulldozer?

Suppose the holes were placed in the center of each side. How would this affect the spin? Where should the holes be placed to make the cartoon rotate in the opposite direction?

The Science

The streams of water produced an equal and opposite force. This force pushed against each of the carton corners. Since the force was applied to the left corner of each side, the carton spun in a clockwise rotation.

If the holes were placed in the center of each side, no spinning would occur. Although four forces would exist, they are positioned opposite to each other and would cancel each other out.

If the holes were placed on the right end of each side, the cartoon would spin in a counterclockwise rotation.

CROSS SECTION

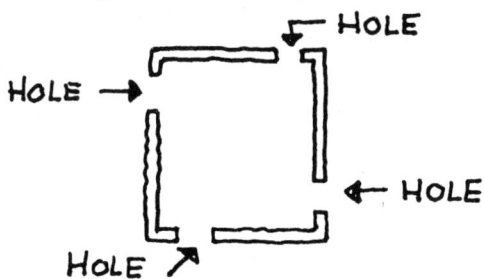

HOLE →

← HOLE

HOLE ↗

← HOLE

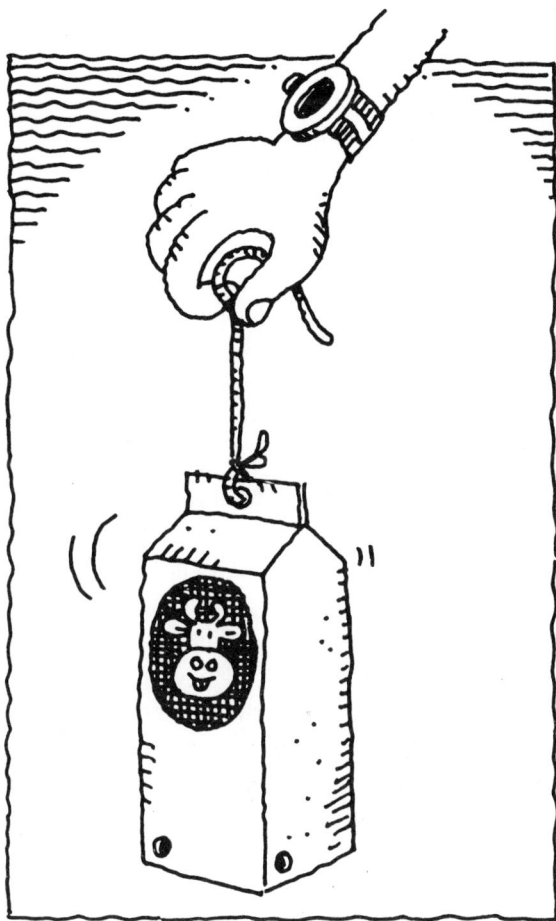

5.3 FLYING POTATO PIECES

-2-1-Zero. Blast off! The potato slice flies into the overhead sky. The crowd roars. Everyone is satisfied, especially the engineers who successfully demonstrated both the laws of physics and the presence of air.

Materials

* *1–2-foot length of 1-inch tube or pipe (made from metal or PVC material)*
* *raw potato*
* *1-foot wooden dowel (less than 1 inch in diameter)*
* *knife*
* *safety goggles*

To Do

Cut several slices of potato. The slices should be larger than the diameter of the tube and about as thick as your thumbnail. Push one end of the tube onto the potato slice. As you push down, the potato should fill the opening to form an airtight plug.

Put on your safety goggles. Turn the pipe to the other end and form a second potato plug. Carefully position a dowel in the middle of the second plug.

When you are ready to launch, push the pipe down. As the pipe slides down over the dowel, the upper potato plug will be launched skyward.

CAUTION

The flying plug can easily damage objects, people, and pets. Make sure that you launch the plug in a safe location. To ensure your safety, you and all nearby observers should wear protective goggles.

The Science

As you lower the pipe onto the dowel, the bottom potato plug rises. As it moves up the pipe, it compresses the air trapped within the tube.

Eventually, the pressure becomes great enough to push the upper potato plug from the pipe. The plug is propelled by a blast of compressed air and "pops" out of the pipe.

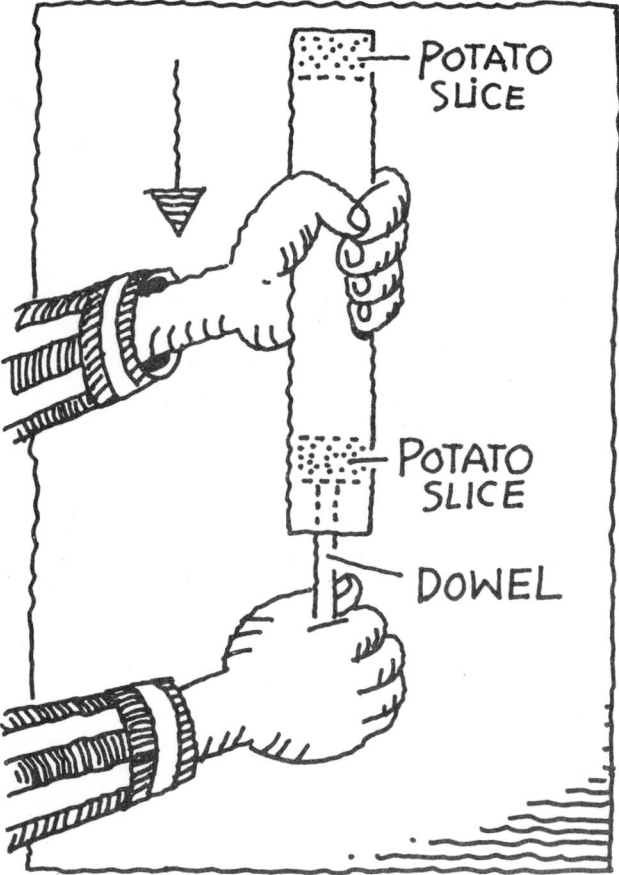

5.4 TWISTED SPOOLS

Rubber bands are made from an elastic material. When this material is stretched, it stores energy. The more twists in the band, the greater the amount of stored energy. When the band is released, the stored energy is given off. The following device harnesses the energy released by an unwinding elastic to power a racing spool.

Materials
- *empty thread spool*
- *metal washer*
- *paper clip*
- *cotton swab*
- *rubber band*
- *tape*

To Do
Insert a rubber band through the center hole of a thread spool. Slide one end of the rubber band over a paper clip. Use tape to secure the clip to the side of the spool.

At the other end, insert the rubber band loop through a metal washer. Then insert a cotton swab into the open loop as shown in the illustration.

Wind the swab so that the rubber band twists. Turn the spool on its side and place it on a flat surface. Release. What happens? How can you control the path of this racer?

The Science
The twists of the rubber band store energy. When the spool is released, these twists are allowed to unwind. The unwinding motion is transferred to the spool through the fixed paper clip. This causes the spool to spin and race along the floor.

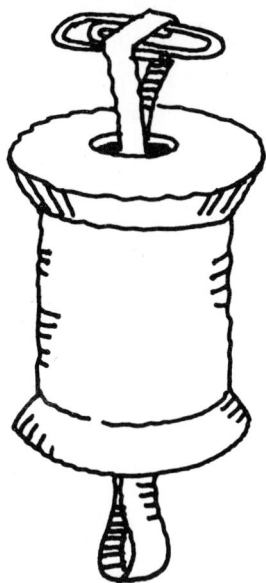

5.5 COME ON BACK, CAN

ere's one of those strange tricks that can baffle your audience. Roll a can and watch as it travels along the floor. Then whistle for it to return. As if responding to the command, the can stops moving and spins back to you. Magic? No, science.

Materials
- *empty coffee can*
- *two plastic coffee can covers*
- *1–5 ounce fishing sinker*
- *metal twist tie*
- *several long rubber bands*
- *nail*

CAUTION
The coffee can has sharp edges. Make sure that an adult assembles the finished can.

To Do
Use the nail to carefully punch two holes into each of the coffee can lids as shown in the illustration. Then cut the rubber bands to form lengths of elastic. Tie the ends of the elastic together to form one strand of rubber about 20 inches long.

Thread the elastic strand through the lid holes as shown in the illustration. Tie the ends of the rubber bands together. Use a twist tie to attach a sinker to the knotted bands.

Have an adult fold one of the lids and push it through the metal container. WATCH OUT FOR SHARP EDGES! Place the lids on the opposite ends of the can. Roll the can and watch as it listens to your "come on back" command.

The Science
As the can rolls, the fishing weight doesn't spin. Instead, it acts as a stationary anchor. As the can spins, the elastics twist up against the fishing weight. By the time the can stops, a considerable amount of energy has been stored in the rubber band twists. When the strand unwinds, the released energy is changed into the motion of the can's return.

5.6 PROP POWER

What do a propeller and a wood screw have in common? Plenty. They are both simple machines that use something called an *inclined plane*. An inclined plane is a simple machine that produces an advantage. Perhaps the most familiar inclined plane is your plain old ramp.

Now imagine that ramp wrapped around a central core. Presto, a screw. Look at the sides of the screw and you'll see the ramp begin at the screw's tip and circle its way back along the screw's shaft. A propeller also uses this type of wrapped ramp. In fact, propellers are known as "air screws."

Materials
* scrap block of foam packing
* two paper clips
* tape
* modeling knife
* kite string
* rubber band
* a toy propeller assembly (removed from a cheap and broken wooden toy)
* safety goggles

To Do
Open and rebend a paper clip so that its ends form two similar loops (clip A). Use tape to anchor this clip to the top of the foam block.

Open and bend a second paper clip (clip B) so that the loop at the back of this piece will hold the propeller assembly. Use tape to attach this clip to the bottom of the foam block. Stretch a rubber band from the front hook to the hook on the propeller shaft.

Tie a kite string across a room. Put on your safety goggles. Wind up the elastic. Hang your craft from the two top hooks. Release the wound propeller and watch your craft shoot along the string!

The Science
As the elastic unwinds, it transfers its motion to the propeller. The spinning propeller cuts through the air. As it rotates, it produces a

substantial flow of air. This flow produces a difference in air pressure that causes the craft to move ahead.

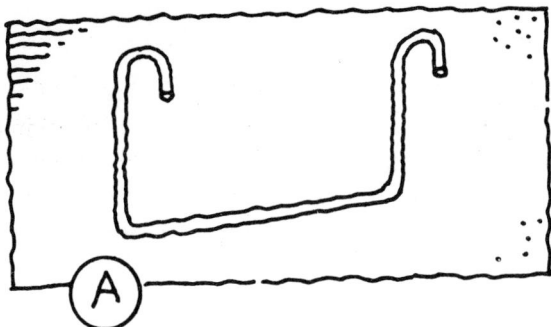

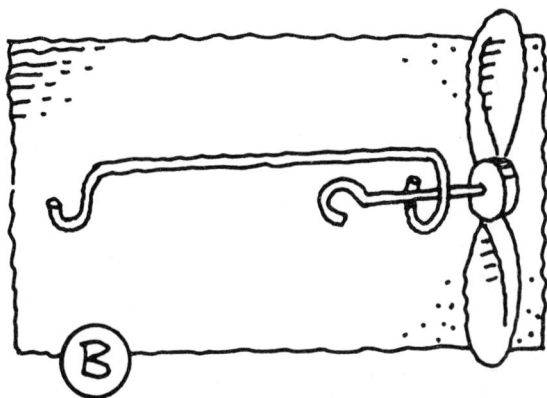

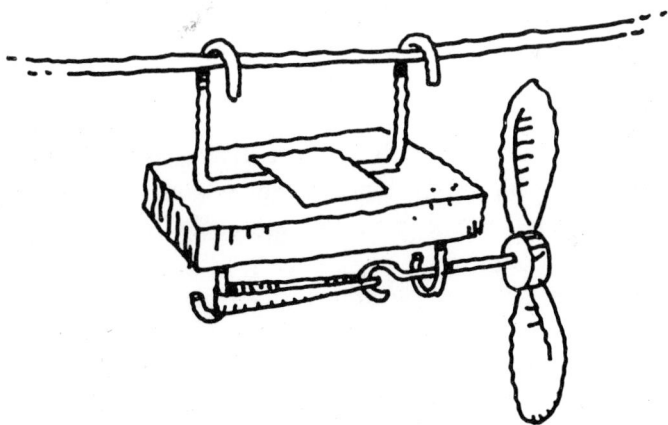

5.7 A REACTION TO AN ACTION

*Like a rubber band, the skin of a balloon is made from an elastic material. When the balloon is inflated, this skin stretches. The stretched skin stores energy. When the balloon deflates, the stored energy is released. By understanding this transformation, you can harness elastic materials to power all sorts of things, including a simple rocket balloon.

Materials
* *several balloons of different shapes and sizes*
* *index cards*
* *pair of scissors*
* *pencil*

To Do
Use your scissors to make two side-by-side snips into index the card. The snips should be about 1 inch apart and placed in the center of the card's shorter side. These snips will form the edges of a bendable tab.

Using a pencil, punch a small hole in the center of this tab. Inflate a balloon. While preventing the air from escaping, slip the neck of the balloon through the hole. Release the balloon. What happens?

Try launching several other balloons. Can you improve the card design to produce a better and more stable rocket?

The Science
When stretched, a balloon skin develops a strong elastic force. This force pressurizes the air trapped within the balloon. When the pressurized air is allowed to escape, it rushes vigorously from the balloon's nozzle.

English physicist Sir Isaac Newton's Third Law of Motion states that "for every action, there is an opposite and equal reaction." In this example, the action is the jet of air escaping from the balloon nozzle. The reaction is the movement of the balloon in the opposite direction from the air jet.

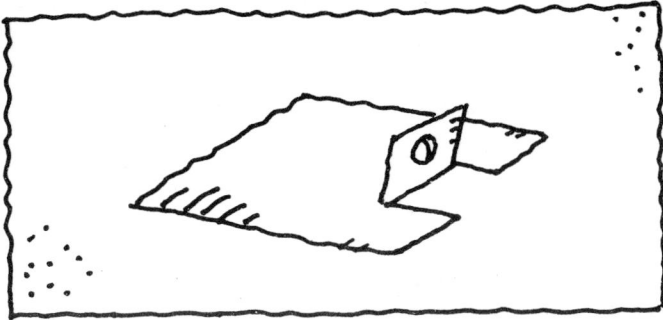

5.8 ON TRACK

May 2013

Now that you know the physics of balloon rockets, it's time to get the most mileage out of this understanding. In the previous experiment, you added stability to your rocket by including an index card stabilizer. In this one, you won't need a stabilizer. Instead, your rocket will be guided on a straight line course by a length of kite string.

Materials
* balloon
* kite string (at least 20 feet long)
* tape
* a couple of drinking straws
* two chairs
* safety goggles

To Do
Pass the string through a straw. Tie the ends of the string to chairs positioned at opposite sides of the room. Separate the chairs so that the string is taunt.

Put on your safety goggles. Inflate a balloon. While holding the balloon nozzle, tape the balloon to the straw. Release the nozzle and watch the balloon rocket across the length of the string.

Try altering the design of the rocket by shortening the straw. Perhaps two straw segments work better than a single straw? You can even control the rate at which air escapes by adding a straw nozzle as shown in the illustration. Keep experimenting with the design until you've built the best rocket, and above all HAVE FUN!!!

The Science
Once again, it's Newton's Third Law of Motion. As the compressed air bursts out of the nozzle, the rocket blasts ahead in the opposite direction.

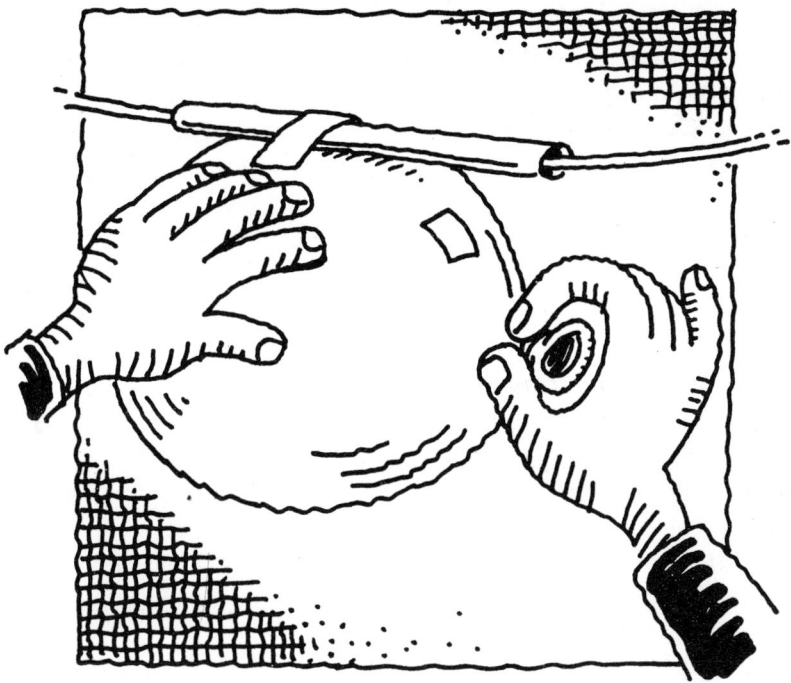

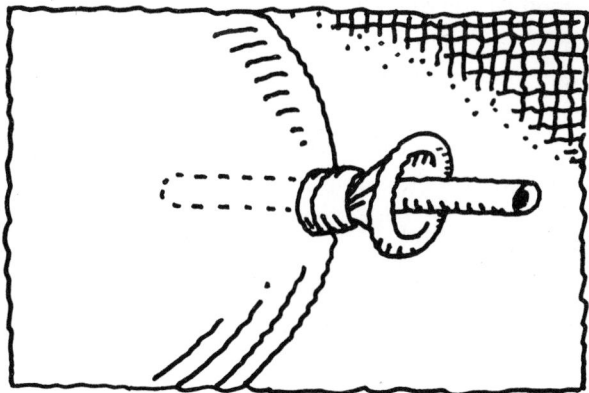

5.9 BALLOON CARS

Now that we've reached the final experiment, this is your chance to design your own balloon-powered racer. Go crazy with the design. Push the envelope. Have fun.

Just think, there are no restrictions. Your balloon-powered cart can have four wheels like a car or it can have three wheels like a tricycle. You can build the wheels out of cardboard, thread spools, bottle tops, or even use wheels that you've ripped off from toy cars. It's up to you.

Materials
* balloon
* tape
* cardboard
* thread spools
* pair of scissors
* glue
* rubber bands
* wooden sticks
* tape
* toy wheels
* straw
* whatever other materials you want to use
* safety goggles

To Do
Think about your design. You may want to brainstorm different ideas so that you won't be locked into a finished "look" right away. Take your time. Try drawing several designs. You may even want to draw up a set of blueprints.

Once you have decided on a design, try building it. Don't worry if you get into problems with its construction. Just figure out a way to get around the obstacle. If there are too many problems with your design, you can always try thinking up a new car. So don't get hung up on what you don't have. Be creative with the materials you do have! And don't forget to wear your safety goggles when you're testing your designs.

The Science

It's Newton's Third Law combined with some creative freedom and mixed with a good dose of heavy-duty brain waves.

INDEX

ANSWERS FOR CHECK IT OUT!

p. 12 the coin is more apt to be "dragged" away; p. 16 compare the spin behavior of a water- and ice-filled container; p. 18 most juggling and balancing tricks; p. 21 going up-weight peaks, going down-weight drops; p. 31 yes, an arc-like shape with no material at the center; p. 37 determine mass of sugar after evaporation of liquid; p. 40 compare the level at which an egg floats in each solution; p. 43 cool air; p. 63 beads on waxed wood, spreads out and soaks into unwaxed wood; p. 72 no; p. 78 basket sinks; p. 83 follows curved path; p. 134 lead cyclist reduces air resistance for other team members.

ABOUT THE AUTHOR

MICHAEL ANTHONY DISPEZIO is a renaissance educator who teaches, writes, and conducts teacher workshops throughout the world. He received an M.A. in biology from Boston University, and for six summers was a research assistant to Nobel laureate Albert Szent-Gyorgyi.

After tiring of counting hairs on copepods, Michael traded the marine science laboratory for the classroom. Over the years, he has taught physics, chemistry, earth science, general science, mathematics, and rock 'n' roll musical theater.

To date, Michael is the author of *Critical Thinking Puzzles*, *Great Critical Thinking Puzzles*, *Challenging Critical Thinking Puzzles*, and *Visual Thinking Puzzles* (all from Sterling). He is also the co-author of eighteen elementary, middle, and high school science textbooks and has been a "hired creative-gun" for clients including The Weather Channel and Children's Television Workshop. He also develops activities for the classroom guides to *Discover* magazine and *Scientific American Frontiers*.

Michael was a contributor to the National Science Teachers Association's Pathways to Science Standards. This document set offers guidelines for moving the national science standards from vision to practice. Michael's work with the NSTA has also included authoring the critically acclaimed NSTA curriculum, *The Science of HIV*.